Radical Political Economy

Radical Political Economy

A Critique

Andrew Sayer

BLACKWELL
Oxford UK & Cambridge USA

First published 1995

Blackwell Publishers Ltd
108 Cowley Road
Oxford OX4 1JF
UK

Blackwell Publishers Inc.
238 Main Street
Cambridge, Massachusetts 02142
USA

British Library Cataloguing in Publication Data
A CIP catalogue record for this book is available from the British Library.

Library of Congress Cataloging-in-Publication Data
Sayer, R. Andrew.
Radical political economy : a critique / Andrew Sayer.
p. cm.
Includes bibliographical references and index.
ISBN 0–631–19374–X. – ISBN 0–631–19375–8 (pbk)
1. Radical economics. I. Title.
HB97.7.S295 1995
330 – dc20   94–23744
CIP

Typeset in Sabon 10 on 12pt
by CentraCet Limited, Cambridge

Printed in Great Britain by T.J. Press Ltd, Padstow, Cornwall

This book is printed on acid-free paper.

Contents

Preface

'Restructuring', 'capital and class', 'uneven development', 'the reproduction of inequality', 'crisis tendencies of capitalism', 'deindustrialization and unemployment' were the kind of key words and phrases that animated large numbers of researchers in the recession of the late 1970s and early 1980s. Now they have a distinctly dated air about them. It is not that their referents have disappeared. On the contrary, they are all too common today: in Britain the current recession is in many ways deeper than that of the early 1980s, social divisions and their human costs have grown continuously; in US cities, the new Germany and many other countries, deepening inequalities and racism feed off one another; in Europe the path to integration is threatened by economic polarization; and at a world scale the North-South gulf continues to deepen. Yet radical political economic theory now arouses much less interest than it once did. Why this should be so is unclear. With the rise of the New Right, the demise of state socialism, and the rise of concerns with modernity rather than capitalism, the reception of Marxist and radical political economy has become, to say the least, sceptical. Yet despite the continuing importance of its subject matter, the scepticism tends to have led to only limited efforts at critique and reformulation, chiefly from 'analytical Marxists'. This book is motivated by the belief that this situation is untenable.

My interest in these matters grew out of earlier work with Kevin Morgan on industrial development (*Microcircuits of Capital*, 1988) and with Dick Walker in *The New Social Economy: Reworking the Division of Labour* (1992). In the former we began to notice systematic silences and reductions in radical political economy, particularly deriving from its attempt to treat most phenomena as effects of class. In the latter we explored how far the neglect of the division of labour accounted for these defects. In the final chapter of *The New Social Economy* we introduced the argument that Marxism's failure to come to terms with the intractability of an advanced division of labour led not only to the failures of centrally planned state socialism, but to a systematic misreading of capitalism. Here, I take that argument further, developing it into a fuller critique of Marxist and other radical accounts of uneven development, and opening up a dialogue between Marxism and liberalism; like certain others on the Left in Britain, I have found Hayek's ideas occasionally fruitful as well as bizarre (Tomlinson, 1990; Wainwright, 1994;).

Two comments are needed regarding the nature of this exercise. First, this is intended as a 'sympathetic' critique, not a hostile one. I believe we can learn from the achievements as well as the limitations of the seventies and eighties boom in research on political economy. As John Stuart Mill pointed out, critiques are useful not only for exposing the weaknesses in our theories, but for reawakening our awareness of their strengths through having to defend them. Otherwise, 'both teachers and learners go to sleep at their post, as soon as there is no enemy in the field' (1859, p. 53). Surprisingly, given the hostile environment provided by the rightward drift of politics in the eighties and nineties, radical political economy is in greater danger of dying through neglect rather than critique, though if it were to die it would be better to do so as a result of critical injury, for we would then at least have some good reasons for its demise. However, I believe that in fact much of it can survive. I have therefore no intention of following in the footsteps of the many well-known converts from Marxism to extreme liberalism. I ask readers to avoid treating my limited endorsements of particular features of socialism and capitalism as indications of wholesale support for those systems, and to resist the temptation to interpret any criticism of Marxism or socialism as an endorsement of liberalism or capitalism, and vice versa. Particularly at this time of disarray in political economy, it is important to acquire the habit of being unafraid to praise and to condemn different aspects of the same doctrines in the same breath. This book should therefore provide something to displease everybody. It is neither a hatchet job nor a rescue job, but an attempt to stimulate a thorough rebuilding of radical political economic theory.

Second, the critique focuses on aspects of radical political economy which have far-reaching implications, but it is nevertheless still only a *selective* critique – relating to the formal economy in advanced economies. It does not pretend to cover all the major issues which need reconsidering in political economy.

Finally, a note on terminology: 'radical political economy' includes but goes beyond Marxism, embracing theorists who diverge from it in many respects, for example those who reject Marx's unqualified resistance to markets. 'Political economy' and 'liberalism' are broad categories with rich histories. By 'political economy' I mean approaches which view the economy as socially and politically embedded and as structured by power relations. Nowadays it is often seen as the preserve of radical researchers, particularly Marxists, neo-Marxists and post-Marxists, though neo-Ricardians, Left-Keynesians and Left-Weberians might be included too.[1] As such, political economy is often counterposed to a more right-wing 'economics' or 'liberal economics', in which economic affairs are treated ahistorically and as largely separable from political and social affairs. Neoclassical, Keynesian and Austrian economics are all variants of this. I realize that this terminology is unsatisfactory in historical terms; eighteenth- and nineteenth-century economic liberals such as Adam Smith and John Stuart Mill regarded their work as political economy, and they did not ignore power or try to abstract economic processes from social organizations to the same extent as their neoclassical successors. Nevertheless, for the purposes of this book I shall adopt the contemporary usage, in which 'political economy' or 'radical political economy' is seen as dominated by Marxist influences, and counterposed to economic liberalism. Where necessary, I shall use more specific terms such as Marxism or Austrian economics.

Acknowledgements

For support and/or comments and disagreements, my thanks to Lizzie Sayer, Hazel Ellerby, John Allen, Kevin Morgan, Doreen Massey, Dick Walker, Michael Storper, Christine Doel, John Davey, Tim Hardin, the Gypsy Kings, Thomas Tallis, Emmelle cycles, Bob Jessop, Beverley Skeggs, John O'Neill, Wendy Olsen, Larry Ray, Balihar Sanghera and my other new Lancaster colleagues.

[1] I realize Marx saw his work as a *critique* of political economy, one which extended to the practices that (bourgeois) political economy both reflected and informed. Political economy in the sense in which we use it here is obviously also a critique in this double sense.

1

Introduction

Having once been highly influential, dominating much of social theory in the 1970s and early 1980s, radical political economy has gradually faded. As the political climate has changed, many researchers have moved into middle-range theory and empirical research, and less economistic concerns. Critical exchanges have taken place between Marxism and feminism, postmodernism and new social movement interests. 'Post-Marxism' – understood as a position that retains some important concepts of Marxism but goes significantly beyond or outside it too – is largely a response to these interchanges. Many of these developments are progressive, but the fading of radical political economy is problematic, for the economic problems which exist now are no less serious than those of its heyday: some kind of radical political economy is still needed. Post-Marxism is much more open and pluralist than Marxism, dropping the latter's totalizing pretensions and sensitive to forces other than those of capital and class (McLennan, 1989), but its stance towards the traditional heart of radical political economy, concerning the formal economy, is unclear. Meanwhile, despite the rightward drift of politics, radical political economists tend to have ignored liberal theory instead of engaging with it. Similarly, despite the epochal significance of the demise of state socialism, it has made little impact on radical political economy, perhaps because of the long-standing gulf between Western Marxists interested in capitalism and those who studied the 'actually existing socialist countries'. A critique is therefore overdue.

The aim of this book is to re-evaluate radical political economy, particularly in terms of its treatment of division of labour and the distribution of knowledge and power in advanced economies. I shall argue that the radical theory of capitalism has lessons to learn both from the problems of state socialism and from liberal theory. Although

Marxism and liberal economic theory are at loggerheads on many fundamental issues, in some cases their strengths and weaknesses are complementary. Furthermore, we find both tensions and striking affinities between liberalism and postmodernism. Radical political economy therefore needs to extend the dialogues which it has already established with other branches of radical social theory, and engage with liberal economic and political theory. For much of the time I will be negotiating between these views, finding faults and insights on all sides, and assessing what needs to be dropped and what can be salvaged.

I believe the most serious outstanding theoretical problems of radical political economy lie in its treatment of division of labour. Marxism acknowledges the historic importance of the rise of division of labour in economic development but assumes that a socialist or communist economy could reach an even higher level of development while overcoming the division of labour. Ironically for a materialist theory, this curious combination of ideas derives from a failure to appreciate the materiality, complexity, opacity and intractability of an advanced division of labour. Unless we recognize these qualities and those of the parallel division of knowledge in society, we cannot hope to understand one of Marx's main targets – the uneven and so-called 'anarchic' character of capitalist development. In the absence of such an appreciation, theorists tend to mistake the fragmented quality of capitalist ownership as a cause rather than an effect of the anarchy of the social division of labour. Similarly, we are liable to mistake the flawed rationality of exchange-value as a consequence of capitalism rather than of the necessity of money in advanced economies (e.g. Collier, 1994a). Equally, we cannot expect to understand the problems of economic motivation and coordination that plagued the state socialist economies unless we recognize the material and informational properties of the division of labour which they were trying to control. Nor, in addition, can we hope to develop feasible alternatives to capitalism and state socialism without appreciating these constraints. Without such an understanding, radical political economy is prone to (mis)attribute the effects of these divisions to other sources, particularly class.

Division of labour is of course associated with early political economy and the work of Adam Smith.[1] It also figured prominently as fundamental to modernity in the writings of social theorists such as Weber, Simmel, Tönnies and Durkheim, though they did not follow through its implications for political economy. In recent decades it has received less attention and most theorists take it for granted or fail to distinguish its

---

[1] It also deserves to be associated with Marx, for it has a more prominent role than class in his early work (Rattansi, 1982).

effects from those of markets. Liberal economists in particular tend to subsume division of labour, indeed the whole economy, under 'the market'. Yet markets are just one of several modes by which divisions of labour can be coordinated; there are also networks, planning, democratic control and conventions. They all have their different strengths and limitations but these depend on the nature of the diverse activities and kinds of information which they have to coordinate. It is also common on both Left and Right to run together markets and capitalist private property, as if no other property relations could co-exist with markets. But state-owned, or cooperatively owned enterprises can also produce for markets, as can petty producers. By extension, markets are not uniquely tied to class relations as sometimes supposed (e.g. Clarke, 1982; McNally, 1993). Whether formal ownership yields actual control over property and activities depends again on the material and informational qualities of their objects. Thus the token character of 'social ownership' derives not merely from contingent forms of organization but from the fact that millions of people cannot hope to control and coordinate the products of property that is diverse, often dependent on arcane specialisms and information, and highly dispersed. The connections between division of labour, markets and property therefore need unpacking. Doing so is helpful in several ways: it helps us to attribute causal responsibility for particular effects within capitalism more accurately, so that we can decide what exactly it is about capitalism which is responsible for them; it helps us work out which elements within capitalist economies are genuinely unique to them and which can also exist outside them, say in market-socialist or state-socialist economies; and lastly, and in consequence of these points, it helps us consider alternatives to capitalism which don't involve abolishing markets or introducing widespread social (i.e. state) ownership.

Although markets should be distinguished from division of labour, they still require particular attention. While they are not as important as liberal economics assumes, they nevertheless provide the main mode of coordination of the global social division of labour. Secondly, the prominence and ideological qualities of the *rhetoric* of 'the market' have been especially strong in recent years, with the market being cast as saviour of the former communist countries and infallible guide of the more liberalized capitalist countries. Thirdly, as is usually the case with terms on which so much is loaded, it conceals several different senses which are frequently confused, leading to incoherent accounts and prescriptions.

In *Capital*, Marx notes that the organization of a factory has a 'double nature', deriving in part from its scale and industrial character, and in part from its 'capitalistic shell'. I shall argue that while this is so, Marx and radical political economy generally fail to note that other aspects of industrial society which they treat as having a single, or uniquely capitalist nature too, also have this dual quality. Thus urbanization, the social division of labour and its anarchic development have a double nature, deriving from the industrial as well as the capitalist character of society. This, of course, has major implications for alternatives to capitalism which assume an advanced industrial economy, suggesting that we should expect many problems merely to change form rather than disappear altogether. The general thrust of the critique therefore leads towards a revived but moderate version of the industrial-society thesis – that is, one which accepts that the social relations of production make a significant difference to the type of industrial society, but which insists that there are nevertheless certain features relating primarily to division of labour that are common to any advanced industrial economy, and merely take different forms under different modes of production. It also makes certain concessions to liberal theory, particularly Hayek's emphasis on the division of knowledge in advanced economies and the 'epistemological' version of the economic problem, and to Weber's insights on rationalization. Nevertheless we argue for a position which differs from those of both of these theorists.

In order to illustrate the theoretical problems in practice, I shall give examples from a particularly strong branch of radical political economy – Marxist and Marxist-influenced analyses of urban and regional development in advanced capitalist societies. This line of research blossomed in the 1970s and 1980s, both applying and developing Marxist concepts in the explanation of subjects such as industrial restructuring and regional decline, the production and financing of the built environment, urban development and housing crises. For a while, radical approaches dominated this field, influencing even those who were not sympathetic to its political standpoint. Nevertheless, it eventually faded as researchers became more interested in middle-range theory of the particular institutional forms of capitalism. Thus 'postFordism' has now come to generate more interest than 'capitalism', though it too will no doubt be succeeded by other concerns. However, we still need abstract theory of the general mechanisms of capitalism and other economic systems, and I shall focus on the accounts of these in the urban and regional literature.

It is widely accepted that social science should 'problematize' social phenomena; even the most mundane and familiar features, indeed especially these, should be treated with wonder and astonishment rather than an indifference bred by familiarity. At times, both Marxists and liberals express this attitude towards capitalism. Marx and Engels' eloquent paean to the revolutionary character of the bourgeoisie in the

Communist Manifesto combines astonishment, admiration and horror. Marxists continue to emphasize the extraordinary transformative power and dynamism of capital to create a world order in its own image. Liberals have a tradition of marvelling at the immensity of the division of labour, the diversity of resources and needs of the millions of people dependent on it, and most of all at the way markets coordinate it without any central agency. These developments are indeed extraordinary. How remarkable that in the space of a few centuries, a large proportion of the world's population should have come to depend on living by producing highly specialized commodities mostly for distant and unknown others, and relying on those others for a huge range of similarly specialized products. Societies which once had only twenty or so major occupations now have tens of thousands; the UK census distinguishes twenty thousand different jobs (Giddens, 1989, 482–3). How remarkable – and dangerous – that our relations to nature, which for so long had been predominantly local or vertical, should be displaced horizontally and specialized to the point where we could imagine that we were not limited by nature.

To marvel at the division of labour and the miracle of the market does not require us to forget about the features of capitalism on which radicals have traditionally focused and accept instead the liberal picture of economies as a network of individuals linked by 'the market', each free to choose what they produce and buy, each independent, self-reliant and free from domination. The coordination of myriad specialized economic activities and demands without a central agency is impressive, but in capitalism it is combined with exploitation, inequalities and uneven development.

Pondering the complexity of modern economies may seem a strange exercise, for we are accustomed to abstracting from complexity in science, indeed the power of theory (if it is powerful) lies in its capacity for simplification. But abstractions give us a grasp of one kind of complexity by abstracting from another. They cut into the connective tissue of the world at different angles, and if we have too few abstractions over too narrow a range of angles we miss important things. It is particularly worth pondering the unfathomable complexity of economies with millions of different kinds of commodities and hundreds of thousands of enterprises, each linked into a global economic order. Even to glimpse this complexity by looking through a product catalogue for just one subsector of the economy, say for computers or clothes, or a 'yellow pages' directory, can be wearying enough. Were it not for the fact that modern economic life depends on the existence of mechanisms which can coordinate this complexity, there would be no harm in abstracting from it in theory. To ignore the complexity of advanced, internationalized economies – as radical political economy generally does – is a disastrous error, one whose effects are all too evident in socialist expectations that such an economy could be controlled by 'the associated producers' through democratic procedures.

Some readers may be sceptical about the possibility of negotiating between radical and liberal views of economies and may consider these to be incommensurable paradigms or discourses. The contrasts are indeed striking. Open any textbook on mainstream economics and one is immediately plunged into exchange, 'the market', choice, the allocation of scarce resources and the relationship between people and commodities. Such starting points, in the space of a paragraph, can simply cut out some of the most important and troublesome issues of political economy. Production and the social relations within it are immediately marginalized. Open a book on Marxist theory, and though you are likely to find more effort in justifying starting points, it will quickly centre on production, its social relations, and the reproduction of economies, thereby disregarding the allocation of resources among competing ends, or micro-economic coordination, which I shall later argue radicals can ill afford to ignore. There could hardly be a better illustration of the power of modes of abstraction to spotlight certain objects while plunging others into darkness. To learn to be a liberal economist or radical political economist is to make these respective modes of abstraction second nature and hence unnoticed. Once this has happened it takes some considerable effort to learn to think in a different mode of abstraction. Yet from these simplified descriptions of radical and liberal modes of abstraction, it should be apparent that there are some complementarities between them – most obviously between the liberal focus on exchange and near-silence on production and the radical focus on production and lack of interest in exchange. Although the political economy/economics split developed in the late nineteenth century with the marginalist revolution, earlier theorists such as Smith and Mill managed to discuss 'horizontal' processes of the allocation of resources and 'vertical' class relations together, and I don't see why we should not be able to do the same. The division between the two approaches has hardened since then, but it is somewhat artificial, not least because liberal economics informs as well as reflects the very practices which radicals study.

Our critique has a methodological side to it. A major problem of any social science, arising from the unavailability of closed (experimental) systems, is the risk of misattribution of causality (Sayer, 1992a), such as the practice already noted of attributing to class what is due to division of labour. It can be highlighted through counterfactual questions and through a critique of abstractions. A good way of assessing causal claims is to ask of any putative explanation what would have happened in the absence of a particular cause or condition, and more generally to conduct throught experiments in which certain changes are introduced into familiar situations. Thus if a certain outcome is attributed to capitalist ownership, it is important to ask how or whether the situation would differ under non-capitalist ownership. I hope to show that the repeated failure to ask such simple questions has allowed radical political economy to be riddled with oversights. Counterfactual inquiries can lead to a 'critique of abstractions', for in following through the implications of these problems we are frequently led to redefine basic concepts or abstractions – in our case, particularly 'division of labour', 'markets' and 'ownership'. Revising any one abstraction by adjusting its sense and reference inevitably disturbs others to which it is related. At first this can cause some conceptual giddiness, but if we are to attribute responsibility or causality properly we must make sure our key abstractions are not burdening their referents with characteristics which should be borne by the referents of other concepts. Our approach is therefore 'disaggregative' or 'de-totalizing' in so far as we seek to work out which elements in any situation are responsible for any given effects (e.g. markets or the particular form of ownership with which they are coupled), which ones are unique to capitalism (or whatever system it is), which ones are common to certain other systems and which ones are separable from the systems in which they appear.[2]

Radical political economy is of course a *critical* social science, both explaining and criticizing the practices it studies, with the explicit aim of reducing illusion and freeing people from domination and unwanted forces. But it can only hope to have an emancipatory effect if it considers its own critical standpoints and the alternative social arrangements they imply. Unfortunately it rarely does this, with the result that its standpoints and implicit alternatives are often contradictory, infeasible, or undesirable even if they are feasible. Marxist-influenced work still bears the traces of the tension between the standpoints of a socialist or communist society which has pre-industrial communitarian qualities and one in which the forces of production are developed beyond current levels of industrialization. More generally, there is a strong modernist tendency in which it is assumed that problems can be progressively unravelled without creating new ones at the same time, as if eventually all trade-offs or dilemmas could be overcome through a triumph of reason. We shall argue through substantive examples that such optimism is not only misplaced but likely to be counterproductive, limiting

[2] This is compatible with a realist approach as defended in Sayer, 1992a.

progress. There are always likely to be 'dilemmas of development' (Toye, 1987).

The problem of critical standpoints has become more acute in recent years, indeed it is central to the crisis of the Left. There is no longer a single standpoint or alternative (socialism/communism) counterposed to a single, overarching target (capitalism). Now there are many targets – patriarchy, racism, homophobia, militarism, industrialism – and correspondingly many critical standpoints with complex relations between them. That critical social science is no longer seen as synonymous with a socialist perspective is a sign of considerable progress, and cause for optimism too, as failure on the traditional front of class politics is compensated by progress on other, newer fronts such as the politics of gender. But it is also a source of heightened uncertainty. While there was always a problem of inconsistencies between critical standpoints, it has deepened and widened with the rise of 'green' concerns, for they bring into question the feasibility and desirability of non-capitalist as well as capitalist industrial societies. Is the problem capitalism, industrial society in general, or modernity?; and what are the alternatives? Equally, increasing awareness of problems of ethnocentrism and value pluralism throws doubt over the familiar, implicit critical standpoints of Western radical social science. How do we decide what is a problem? What if we cannot reach a consensus on this? Until recently, it seemed that the problems or targets of critical social science could be relied upon to emerge from the investigation of existing practices, where one would encounter the felt needs, frustrations and suffering of actors, and in discovering the sources of these problems, work out what changes would lead towards emancipation (e.g. Fay, 1975, 1987; Collier, 1994b). This was coupled with an implicit view that emancipation was a form of escape from domination, illusion and unwanted constraints, with little or no acknowledgement that it depended on the construction of superior, alternative, progressive frameworks which could replace the old ones. But it is now increasingly apparent that normative questions of possible alternatives and what is good or bad about them cannot be evaded. How, without addressing such questions, could one decide what constitutes a superior alternative? Should there be a presumption in favour of community as a basis of social organization over other forms? Does liberalism provide the best framework for multicultural societies? What should be people's rights and responsibilities? What are our responsibilities to distant others, future generations, and to other species? There is little hope of achieving the goal of an emancipatory social science if it shuns normative discussions of issues such as these.

Readers won't be surprised that I don't produce answers to these normative questions (!), but what I shall do instead is sketch out some of the debates in normative political theory and philosophy which bear upon the development of better critical standpoints. Here the tensions between liberalism, Marxism, feminism and theories of difference, and communitarianism are indicative of the kinds of issues which I believe radical political economy needs to take on board if it is to have any emancipatory influence. Although there has been much innovation in political philosophy and theory in recent years, the old issues of equality, liberty, rights and responsibilities cannot be escaped. The development of multi-ethnic societies and the increasing salience of difference may make a logic of fragmentation seem compelling – until, that is, we remember the interdependencies between the fragments via the global division of labour. Thus, while these developments make the task of finding an appropriate common social framework more difficult, they also make it more urgent.

Situating the Crisis of Radical Political Economy

The content of this book has been shaped by a particular view of the development of radical social theory over the last twenty years. It may help, therefore, to set the scene for the critique by sketching out this view.

The 1970s and early 1980s were a time of unprecedented strength for radical political economy in academia. Since then a number of changes have taken place, some through internal debates and developments in adjacent areas of radical theory, others in reaction to external social change and political movements. Not all the changes arose purely through argument; there were also the usual factors of boredom with old concepts, collective amnesia, new interests such as the hitherto neglected field of cultural studies, and a slow shift in dispositions in which certain concepts lose their resonance and become strangely dated, while others come to the fore. Within radical political economy itself, dissatisfaction with the labour theory of value has grown, though there is still no consensus on its fate (Mirowski, 1989; Cohen, 1988). Developments in so-called 'analytical' or 'rational-choice Marxism' have dissected Marxist theory into component claims and subjected them to close scrutiny with striking effects, most notably in questioning the theory of exploitation (Carling, 1986, 1992. Elster, 1985; Roemer, 1986). Whatever doubts might be held about its methodological individualism and its casual approach to the status of its abstractions and premises, analytical Marxism has a refreshingly detached attitude to the heretical implications of its critiques of Marxism. Outside radical academia, critiques striking more directly at the heart of Marxist political economy have continued to come from a much older source – liberalism (e.g. Arnold, 1990; Hayek, 1988). However, my suspicion is that interest in Marxist political economy has weakened less as a result of these internal and external criticisms than through exhaustion or boredom with its internal debates at a time when other developments in radical social science looked more interesting.

One of these innovations was a shift towards 'middle-range theory' and empirical research. Early efforts at applying Marxist concepts tended to be reductionist, underestimating the range of constitutive processes and outcomes (Sayer, 1981, 1985; Storper, 1985). It quickly became clear that the more abstract political economic theory needed supplementing by middle-range theory to make sense of these different forms. For example, as the diversity of types of labour process was recognized, radical researchers moved from Marx and Braverman (1974) into studies which incorporated insights from other theories (e.g. Littler and Salaman, 1984). Although capital is capital wherever it is, there are vast differences between, say, Japanese capital and British capital, between Fordism and post-Fordism, and so on. Similarly, students of capitalist urbanization could hardly fail to note the huge differences between, say, Paris and Los Angeles, and hence they moved beyond early general theory such as that of Castells (1976) into studies of particular institutional forms in urbanization. Typically, middle-range theories deal with the specific forms and mediations of capitalist processes, such as the nature of institutions, or new forms of organization such as post-Fordism. While this has been immensely fruitful it has also been associated with a neglect and dilution of abstract political economy so that it has often become more a source of 'atmospherics' and political dispositions than a self-consciously applied analytical framework (Harvey and Scott, 1987). It almost seemed as if one could dispense with the more abstract theory of political economy and forget questions about the basic social organization of economies. However, this is a view which I shall challenge, for important though middle-range theory is, it can only be expected to supplement and modify rather than supplant more abstract theory. It does not problematize the broad structures of society, such as those concerning the nature and legitimacy of the social organization of the economy – phenomena such as division of labour, property rights, hierarchy, and exploitation. Abstract political economic theories remain invaluable for discussing such matters, even though they need to be qualified via middle-range theory where they are applied to concrete situations.

A second development which drew attention away from the familiar concerns of radical political economy has been the *diversification* of radical concerns since the late 1970s. Radicals began to look beyond production in the formal economy, capital and class as the traditionally prioritized sites of power and domination. Feminist, anti-racist, post-colonialist, green, and other 'new social movements' have had a considerable and salutary impact. Thus gender, race, nationalism, sexuality and differences of identity and culture are now assuming more importance in radical literature. In part these call not only for significant shifts in interest but for changes to the content of traditional radical political economy, particularly in its concrete studies. Thus patriarchy and capitalism (and patriarchy and state socialism) have clearly adjusted to one another in numerous ways. This raises the question of whether they should be considered as a single system or as two interacting systems; has patriarchy influenced capitalism to such an extent that the latter could not possibly function without the former, or could they exist independently of one another? (Delphy and Leonard, 1992; Eisenstein, 1979; Hartmann, 1979; Walby, 1986) (I shall comment on these issues in chapter 3). One of the most general failings which the exposure to these new issues revealed was the 'class reductionism' of Marxist and other radical political economy. By this I mean both the assumption that class divisions and domination were more important politically than others, and the tendency to assume that all problems – such as racism and sexism – derived from the diagnostic characteristics of capitalism, particularly the capital–labour relation, or at least would only be resolved once capital had been overthrown. I have no doubt that these assumptions are absurd, and in so far as I discuss capital and class, I make no such judgements; class is just one among several forms of social division and domination. But there is a further kind of class reductionism which is ignored by feminist and other critiques, and this is the tendency to attribute to class, effects which stem from divisions of labour within and between enterprises. Both kinds of reductionism need combatting.

These two developments – the rise of middle-range theory and the diversification of radical interests – are both needed in radical social science, but it should be possible to pursue them without simply dropping the interests of more traditional political economy, as if class, capital accumulation and uneven development were no longer important. However, further influences have reinforced the drift away from these subjects. One is a certain deradicalization of work on the formal economy, so that implicitly it has become more reformist. Radical political economy slowly shifted from making critiques at the level of mode of production to more specific criticisms of practices within capitalism. For example, studies of industry and uneven development have shifted from a critique of capitalism to analyses of capital's latest developments whose normative implications border on boosterism.[3] This was presumably a difficult transition, for one of the most striking features of the earlier rise of Marxism from the late 1960s was an emphasis on the structural character of problems within capitalism, as against the liberal tendency to treat them as aberrations capable of solution within the parameters of the system. Given the drift to the Right in politics, or perhaps we should say the decline of the Left,[4] this slow but marked change is less surprising, though what is peculiar is that it should have taken place with so little debate and justification. It is the product of a tacit, consensual drift rather than argument. Whatever the reasons for this deradicalization and the manner in which it has come about, two points should be noted; firstly that diversification and deradicalization are separate tendencies – concerns with gender or race are no less radical than a concern with class (indeed they are related to a new politicization of these areas of social life); secondly, that the term 'reformist' is not intended to have any pejorative connotations.

A further contextual influence deriving mainly from academic sources was postmodernism. This has had varying degrees of influence in different parts of academia – strong in sociology and literature, weak in economics, intermediate in politics and human geography. Where radical social theory in the 1970s tended to relate particularities to capitalism, it now situates them in relation to 'modernity' or 'late modernity'. Two main features which encouraged the drift from political economy were postmodernism's overwhelmingly cultural content – it had virtually nothing to say about economies[5] – and its antagonistic attitude towards totalizing theory or grand narratives. In the latter case, Marxism was the prime target. While any reduction in the dogmatism associated with totalizing approaches is welcome, there are also dangers in simply dropping any theory which claims to explain extensive and interlinked systems. In so far as social structures are big and interlinked, approaches which divide and fragment them are likely to mystify rather than illuminate (Sayer, 1993). This is not a covert way of defending orthodox Marxism, for capital and class are not the only big and pervasive systems of power; patriarchy is pervasive too (Mies, 1986; Hartsock, 1990). At worst, whole swathes of theory are in danger of being dumped on the dogmatic and *a priori* grounds that they constitute a totalizing discourse, which of course saves anyone the hard work of critically engaging with

---

[3] Witness the researches following Piore and Sabel's *The Second Industrial Divide* (1984).

[4] In Britain, the academic Left was relatively insulated from the hostile political environment, having consolidated its position sufficiently for many to remain in post and in receipt of research grants.

[5] Nevertheless, Harvey has made a strong case for regarding postmodernism as arising largely in response to unacknowledged political-economic developments (Harvey, 1989).

their substantive content. Whether such a discourse excludes important areas of social life while purporting to be all-inclusive is emphatically an a posteriori matter. There is no alternative to sifting through available theory to see what is adequate and what should be abandoned or changed.

The most dramatic external political events were the revolutions of 1989 and the subsequent fall of the Soviet Union. For the Right they were a vindication: communism was defeated, capitalism had won. The 'end of history' was announced (Fukayama, 1992) and 'market triumphalism' swept the board. Strangely, it was assumed by many that it also meant that Marxism was defeated, as if Western Marxists ought to have recanted as soon as the statues of Lenin and Marx started being toppled. This ignored the fact that the vast majority of Marx's work and most Western Marxism is concerned with *capitalism* and that most Western Marxists were overwhelmingly critical of the state socialist regimes. The obvious response for western Marxists was put most succinctly by Frederick Jameson: 'it does not seem to make much sense to talk about the bankruptcy of Marxism, when Marxism is very precisely the science and the study of just that capitalism whose global triumph is affirmed in talk of Marxism's demise' (1991, p. 255). Not understanding or accepting this, the Right found it hard to believe that Western Marxists could actually have celebrated the velvet revolutions too. As a Marxist who wanted 'neither Washington nor Moscow', Alex Callinicos could rejoice in the death of the state socialist countries (not his term), since it cleared the political ground of a diversion from the central task of dealing with the 'unfinished business' of attacking capitalism (Callinicos, 1991).

Yet while the 'velvet' character of the revolutions was remarkable enough, there was little else that the Left could celebrate about them. As Habermas points out, they were also singularly depressing in that they were devoid of 'ideas that are either innovative or orientated to the future' (1991, p. 27). Whether Habermas meant it or not, I would add that it was Western Marxists as well as people in the former socialist states who lacked ideas about alternatives. In this context, market triumphalism could divert attention from the continued failings of capitalism, as if the 'victory' of capitalism meant that no one had any right to criticize it. Again, as Habermas put it, 'it is not as though the collapse of the Berlin Wall solved a single one of the problems specific to our system' (Habermas, 1991, p. xii). While the latter statement is surely correct it could be read as implying that it was 'business-as-usual' for the Left. It is my view that *this* kind of interpretation, together with those of Jameson and Callinicos, are complacent and hopelessly inadequate. One can agree with Jameson that Marxism is primarily a theory of capitalism, but this position is nevertheless all too smug, for it begs the question of whether its account of capitalism is at all adequate.[6] Similarly, Callinicos implies that there are no lessons to be learned from the demise of state socialism, save that it wasn't real socialism, and there are certainly no lessons for the critique of capitalism.

This book is motivated by the view that such complacency is entirely unwarranted. The totalitarian character of state socialism and its problems of economic motivation and coordination are not historical aberrations but are presaged by Marxism's lack of a sufficiently materialist understanding of the social division of labour and its associated division and dispersion of knowledge in advanced economies. This failing not only explains the inadequacies of state socialism's attempt to plan such an economy centrally, but is the major unresolved flaw in Marxist theory of capitalism. The reluctance of the Left to think through alternatives (for fear of producing 'blueprints' which might pre-empt future struggles) meant not only that radical political movements had little idea of feasible and desirable objectives, but that the standpoints from which capitalism and its problems were explained and criticized were unexamined and often incoherent or undesirable. There is no way the Left can reply to market triumphalism and the lack of alternatives without giving some consideration to the old problems of political economy.

Out of all these influences and tendencies has emerged a loose 'post-Marxist consensus'. This may seem a strange term because simply naming it thus is liable to provoke a controversy as to what extent it goes beyond Marxism, how much of Marxism it drops, and what else it admits. But many researchers, particularly those engaged in empirical work, have not argued through their position explicitly, and it is their implicit views which constitute the consensus. Its most important characteristics are the usage of markedly diluted and 'de-totalized' versions of Marxist concepts, and an increasing eclecticism. It abandons the privileged status of class, the labour theory of value and the old-time religion of alienation, de-prioritizes the existential qualities of labour, adopts a more positive attitude to consumption (e.g. Miller, 1989), and admits with varying degrees of openness, the importance of non-class sources of oppression and difference, most particularly relating to patriarchy (e.g. Bowles and Gintis, 1986; Carling, 1992; Corbridge, 1993; Elster, 1985; Laclau and Mouffe, 1985; Mouzelis, 1990; McLennan, 1989; Osborne, 1991; Soper, 1991; Slater, 1992; Wright et al., 1992).

A charitable view of the changes in radical social science during the last decade would be that with the rise of new social movements and the

---

[6] The rest of Jameson's essay weaves round a range of issues with consummate complacency.

politics of difference, and in response to the previous gross neglect of these issues and of culture, researchers shifted into the analysis of forms of domination and subordination that went beyond those of capital and class. This seems an entirely reasonable response, as analysis of these matters was long overdue. The uncharitable explanation is that faced with the challenge of the New Right and the weakening of Marxism, radicals shifted their attention to new concerns which did not require them to make any painful concessions to the Right. Socialism as an alternative political system was increasingly difficult to articulate and defend, but in any case there were other important issues to turn to which provided a convenient escape. While understandable, this escapist response is surely less defensible. In my view, both of these explanations of the shift from economy to culture – charitable and uncharitable – are right. Although post-Marxism has been generally a progressive development in terms of broadening radical interests it has neglected political-economic theory. There is therefore much unfinished business.

Plan of the Book

As will already be apparent, in part, my criticisms of radical political economy are directed at its method and ontology, as well as its substantive content. Chapter 2 addresses these issues in order to clear the ground for the substantive critique. Here I discuss whether 'anti-essentialist' arguments undermine the project of abstract political economy and allow middle-range theory to take its place, as appears to have happened in some quarters. I also argue for a 'disaggregative' approach and an examination of the standpoints from which critical explanations are made, so as to see if they are feasible and desirable.

Chapter 3 contains the central component of my critique – the treatment of divisions of labour in radical political economy. It starts by distinguishing division of labour from class and other social divisions, arguing that they are frequently confused. It then addresses the complexity and intractability of an advanced economy and its implications for capitalist, state-socialist and market-socialist economies. Two convergent lines of argument are then pursued – one concerning the effects of economies and diseconomies of scale and scope on economic power, the other deriving from Hayek on the division of knowledge and the way in which ignoring this encourages an excessive rationalism or 'constructivism'. Together these arguments imply the need to rethink the central distinction between technical and social divisions of labour, and to acknowledge the inevitable fragmentation and dispersion of power in advanced economies.

In chapter 4 we turn to the *modes of coordination* of divisions of labour and their respective properties and limitations. Markets, planning, networks and democratic control are the main modes considered, but special attention is given to markets as they have figured so prominently – often without illumination – in debates within and between radical political economy and liberal economics. In order to expose the way in which people use quite different concepts of the market, often sliding between them unknowingly and arguing at cross-purposes, I include a conceptual analysis of these different meanings. This shows that while most meanings have a reasonable 'home' context in which they are useful, they are frequently over-extended and expected to be informative in other contexts.

Having clarified some of these conceptual differences we turn in chapter 5 to some of the issues raised in these debates on markets. Many involve negotiating between liberal and radical approaches. The invisible-hand and consumer sovereignty theses associated with liberal economics are only partly compromised by the consideration of issues of power and the social embedding of markets highlighted by radical political economy. The liberal concern with allocational efficiency is typically either dismissed or not understood on the Left. We argue that while it is an important issue for any kind of economy, and one which the Left needs to consider with respect to both capitalism and socialism, its significance needs to be reconsidered in relation to efficiency in production and innovation – which radical political economy understands better than liberal economics. We then evaluate how far markets have egalitarian or inegalitarian effects, and whether they are neutral with respect to differences and inequalities such as those of gender and race. Here we find several different mechanisms, sometimes working in different directions, but in all cases characterized by amoral incentives: where markets are concerned, the key question is 'will it pay?' Finally, we consider the Panglossian assumption – that is, that the existence of competition is sufficient to ensure that outcomes will be roughly optimal even if actors are not always optimizers. Here we find that the notion of the survival of the fittest relies upon a seriously inadequate view of evolutionary processes which exaggerates both the benefits of competition and the dangers of intervention.

In accordance with our disaggregative approach we then turn in chapter 6 to ownership and control – relating this not only to markets, but more fundamentally to divisions of labour. We examine the relationships between specialized activities and the information and knowledge associated with them, and how these influence and are influenced by ownership and control. In chapter 7 we outline and compare ideal types of three non-capitalist economic systems – petty commodity production, market socialism and state socialism. These provide counterfactuals and comparative material for understanding and evaluating capitalism, helping to clarify both what is distinctive about it and what it shares with other economic systems.

Chapter 8 applies the critique to radical theories of urban and regional development, particularly to the work of Harvey, Scott, Lojkine, and Eisenschitz and Gough on capitalist industrial restructuring, uneven spatial development and urbanization. In several different ways, we find dilemmas of development that are likely to be found in any advanced industrial society, being treated as unique products of its capitalist form.

The concluding chapter attempts to draw out the implications of the critique, firstly by positioning it in relation to other theoretical tendencies. Thus we return again to the tensions and complimentarities between radical and liberal modes of abstraction, and focus on the debate between their contrasting 'substantive' and 'formal' conceptions of the economic problem. We clarify how our critique relates to older versions of the industrial society thesis and to the work of Weber, Foucault and Hayek. The relationship to postmodernism is discussed via its largely unacknowledged affinities with liberalism. Secondly, and following up our earlier arguments for an examination of the critical standpoints of radical theory, we explore the relationship between critical social sciences such as radical political economy and normative theory. We conclude by reviewing what debates in political theory and philosophy have to offer the search for emancipation.

2

Questions of Method: Abstract Theory, Counterfactuals and Critical Standpoints

One cannot expect to progress far in assessing political economy without getting involved in methodological and philosophical debates about approaches to the subject. Although, as I suggested in the first chapter, the changes in the nature and status of radical political economy are almost certainly reactions to an increasingly hostile political climate, they are also the product of ongoing debates about approach. In this chapter I shall consider three areas of controversy, both actual or potential, in order to prepare the ground for the substantive critique of the subsequent chapters.

The first concerns the status of abstract theory, relative to empirical work and 'middle-range theory'. This is important if we are to evaluate whether the shift out of abstract radical political theory is justified on methodological and theoretical grounds, or whether it is merely a consequence of changing academic fashion and weakening commitment to the political associations of that theory. In taking up this question we encounter the 'anti-essentialist' critique, with its rejection of attempts to discover relatively enduring and widespread essential features of economies, and its implication that middle-range theory must supplant rather than just supplement abstract theory.

Secondly, we assess the extent to which an analytical or disaggregative approach can usefully be employed in political economy, in contrast to the totalizing approach traditionally associated with Marxism. This hinges on fundamental ontological issues regarding whether or how far phenomena are internally related. If all phenomena are internally related, this might imply that everything in a capitalist society is affected by its capitalist nature to such an extent that we cannot consider what any of its social structures or practices might be like in a non-capitalist context. If, on the other hand, society is not a seamless whole in which everything is related to everything else, we can consider how far specific social forms (e.g. certain kinds of property relations) can exist apart from other forms (e.g. modes of micro-economic coordination such as markets) with which they may be commonly, but contingently, associated. In particular, I shall discuss the use of counterfactual questions and thought experiments as ways of assessing the explanations given by radical political economy.

Thirdly, I shall evaluate the critical dimension of radical political economy, since its primary constituent, Marxism, is generally taken to be a paradigm case of critical social science. Critical social sciences not only attempt to explain social processes but provide a critique of them. Unlike natural phenomena, social phenomena depend in part on the understandings, norms and rules that actors use in their everyday practice, and these can be the object of critiques. The discussion focuses on a neglected aspect of critical social science – the 'critical standpoints' from which critiques are made, and the kinds of alternative forms of social organization they imply. I then argue that the critical standpoints of Marx's critique of capitalism are incoherent and that this destabilizes the foundations of his analysis.

Finally I bring the explanatory and critical sides of critical social science back together and look at the relationship between explanations, counterfactuals, critical standpoints and alternatives. As before, my aim is not to abandon radical political economy's critical intent, but to suggest ways of making it more effective.

Abstract Theory and the Anti-Essentialist Critique

The most familiar theories of political economy or economics are highly abstract, that is they are deliberately one-sided, isolating and illuminating particular structures or relationships by holding off contingencies that generally accompany them in concrete situations. By contrast, concrete analysis of actual political-economic phenomena treats them as many-sided, and often draws upon more than one abstract theory. With its extraordinarily insular orthodoxy, liberal economics has not given much consideration to criticism of its abstract character. In radical political economy the situation has been different. Though immensely popular in the 1970s and early 1980s, abstract theory tends to have been displaced by middle-range theory and empirical research since that time. Where previously many researchers felt content with levels of abstraction equivalent to those of Marx's *Capital*, there is now a greater reluctance to abstract from the particular institutional forms in which economies develop. How far was this shift justified? Is there still a place for abstract theory in political economy?

In part the shift was a justifiable response to widespread misuses of abstract political economy in the form of reductionism. By this I mean the reduction of the concrete or many-sided nature of social phenomena to those aspects isolated in abstract theory, as if nothing else mattered or was needed in the explanation of those concrete phenomena. Thus, Los Angeles and Paris might seem remarkably different, but all that mattered was that they were both capitalist cities. Sometimes reductionism works in the opposite direction, when particular concrete forms are 'read off' wholly from the claims of abstract theory (Sayer, 1985; 1992a, pp. 238–9). Thus, a concrete phenomenon such as deskilling of workers had to be interpreted as derivable from abstract political-economic theory alone, rather than from that theory in conjunction with analyses (theoretically informed but by different theories) of the other constituent processes which were jointly responsible for the deskilling. These other elements – such as the state of provision of training or the particular nature of contemporary technological change – tended to be excluded because they were not theorized by Marxism, but by other theories. In a manner earlier criticized by Sartre (1963) and Williams (1962), this kind of 'research' found only what it already knew and learnt nothing. A proper concrete analysis would have attempted to integrate relevant abstract political-economic theory with those other theories so as to comprehend the many-sided nature of the object. In so doing it would have had to combine abstract and middle-range theory. Note that while the language of abstract and concrete is associated with radical political economy, similar problems occur in liberal economic theory, indeed economists attempt to make a virtue out of reductionism.

As the myopic nature of reductionism dawned on researchers in the late seventies and early eighties, many – myself included – moved into empirical research and middle-range theory as a corrective (Sayer, 1985). So strong did this trend become, that by the end of the eighties, abstract radical political-economic theory was becoming neglected. The result now is that instead of having a middle-range theory which rests on the shoulders of abstract theory, it has largely lost touch with it altogether. Certain critics of this tendency complained with some justification that the middle rangers had virtually abandoned abstract theory (Harvey and Scott, 1987). With less justification, they denied that the middle rangers' work had any theory, regarding Marxism as the only theory and anything else as empiricism (e.g. Harvey, 1987). Consequently, their efforts to save abstract theory seemed to call for a rejection of middle-range studies and a return to reductionism rather than a reconciliation with middle-range theory.

While some political economists protested about the desertion of abstract theory, few of the deserters bothered to justify it. There is, however, a parallel development which could be construed as a rationalization of the shift: 'anti-essentialism' (e.g. Laclau and Mouffe, 1985; Hindess, 1987; Tomlinson, 1990).[1] Abstract political-economic theory is characterized by a kind of 'essentialism', which some regard as a virtue and others as a liability. Essentialists claim that objects, whether social or natural, have particular ways of acting as a consequence of their structure or intrinsic nature. Thus, for example, markets tend to distribute resources towards the richest consumers and to penalize inefficient producers, in virtue of their competitive nature and the presence of purchasing power as the sole criterion for the allocation of resources. This kind of essentialism, which is compatible with critical realism, is quite different from what might be termed epistemological essentialism, or foundationalism, which assumes that to identify an essence of something is to claim to have found the ultimate truth about it.[2] After the rise of fallibilism in the philosophy of science in the last three decades, no one seriously believes this any more (though postmodernists like to suppose their enemies still do), and contemporary defenders of critical realism and essentialism are insistent fallibilists, making no foundationalist claims about their knowledge of essences. Their essentialism, then, is ontological rather than epistemological.

In claiming that phenomena such as markets have certain essential features, we do not have to suppose that they always produce the same effects, whatever the circumstances. Rather there are two conditions controlling outcomes, internal (for example, the nature of the institution) and external (the conjuncture in which it is situated). This point can be explained by reference to critical realist philosophy (Bhaskar, 1989; Sayer, 1992a) or Aristotelian principles (O'Neill, forthcoming). First of all, concerning the internal condition, it is contingent whether particular social phenomena come into existence, since they depend on social relations and people's actions. Where they do exist, not all their (essential) powers may be activated at any particular time. A state does not have to be exercising its monopoly of legitimate violence at every moment to be a state, but it has the power to do so. A worker does not have to be working all the time to be a worker. We can also acknowledge that the essence of a social object can, like the meaning of marriage, or

---

[1] Anti-essentialism is a distinctive feature of postmodernist thought. For example, it has been influential in feminism, where the concept of 'woman' as having a single essence has been widely criticized (Grant, 1993).

[2] In my *Method in Social Science*, I attack this and certain other doctrines that have been labelled 'essentialist', and therefore reject the term. However, other uses of the term as described here avoid the problems of those doctrines and seem to me to be compatible with critical realism.

the skills of a worker, change over time, so that essences are historically specific and may be constructed relationally rather than in isolation. It is further an empirical question just how many different kinds of essence there are; the properties of the markets for labour power, capital and products may be sufficiently different for us to decide to define three essential kinds of market rather than just one. Secondly, concerning the external conditions, where causal powers are activated, their effects depend on the nature of their contexts, that is, the causal powers of the other objects constituting those contexts. Consequently, one can argue that an economic institution has certain essential powers, which are dependent on the contingent matter of the institution's reproduction, and that its effects will be mediated by contingently related contexts.

Anti-essentialists also stress the variety of actual behaviours that can occur but conclude that this means that they have no essences. What happens depends on how actors interpret situations and rules and on a host of contextual factors. One of the most prominent anti-essentialists, Hindess, offers another angle on this, interpreting essentialism as a projection of a discursive phenomenon – 'principles' – onto social practices (1987, p. 7). He repeatedly argues that social forms are not the expression of principles, but the complex outcome of material processes. At one level in so far as social practices are concept dependent, Hindess is surely incorrect, for it may indeed be the case that actions are informed by principles – for example, a doctor's by her ethical and professional principles. On the other hand, practices are not reducible to actors' understandings, whether they involve principles or something else, for contingencies, opportunism, and tacit skills and knowledge generally loom large, so that complex processes, such as the behaviour of a large public organization, cannot be read off from or seen as the instantiation of principles. Another way of putting this is in terms of the critique of the *structure–conduct–performance paradigm* (Auerbach, 1988). In this paradigm, which is present in both liberal and Marxist economics, one can read off actors' behaviour from particular situations (e.g. behaviour under competitive as opposed to monopoly conditions) and from this predict performance. This is effectively a form of reductionism, and has been condemned, again on the grounds that actors can interpret a given situation in different ways and hence produce different results. However, not just any interpretation or action will be equally successful. The openness of social systems, and the possibility for varied and novel responses to the same situation does not mean that there is no point in seeking theories of the essential, relatively enduring features of property relations, production, exchange, exploitation, etc., as if all that were needed was middle-range theory and empirical studies of the myriad forms of concrete economies.

Inasmuch as these critiques recognize the daunting complexity of concrete processes, and reject the reductionist goal of reading off concrete events wholly from abstract theory, they are effective, and indeed, we shall use similar arguments in the following chapters. Thus while economic structures constrain and enable action, concrete behaviours can't be deduced from them alone. On the other hand, in so far as anti-essentialism denies the existence of any essences or causal powers, it fails, for this would ultimately imply that anything can happen in any situation. If social objects have no essences – no properties, no causal powers and liabilities – then it is difficult to see how they could do anything, and hence how social life itself could be possible. While we can indeed interpret situations in various ways and act accordingly, we are also constrained or disciplined by existing structures to act within certain ranges. Attempts to act outside these ranges are likely to fail, which is why structures – such as markets – are not merely constraining but enabling, allowing developments which would otherwise be impossible. To be sure, social structures are themselves recursively reproduced and transformed by actions, but since structures are the medium as well as the product of those actions, the latter never break free of structures of some kind, though they may break free of particular ones (Bhaskar, 1989; Giddens, 1984). The danger of anti-essentialism is that it switches straight from determinism and reductionism to voluntarism. Extreme versions of anti-essentialism which suppose that anything can happen in any situation therefore render explanation impossible, for there is nothing that theory can say about what determines what. Extraordinarily, some anti-essentialists actually embrace this denial of causal explanation, arguing that nothing is more influential in producing a given change than anything else (e.g. Graham, 1990), implying absurdly, for example, that the government could as easily be brought down by changing our shoes as by revolution. To explain a complex process no less than a simple one, we have to say what mechanisms (causes including reasons) co-produced it and in virtue of what structures those mechanisms exist. There is no point in an anti-essentialist even mentioning all the possible factors in a concrete situation if they were not responsible for determining what happened.

The fact that at least some abstract theories have widespread applicability across a range of different contexts suggests not only that there are some essential features of economic phenomena but that they are widely replicated. Indeed, remembering that assumptions about practices are used by lay-actors as well as by 'theorists' and are a condition of the possibility of forms of social life of any durability, then though socially created, many social forms can be said to have relatively stable essences. Many social structures or practices can retain their identity across large regions of time and space (e.g. commodity exchange), indeed they have to; processes of 'globalization' depend on it.

To isolate the enduring properties we generally need a more abstract (that is, one-sided or selective) analysis. To identify variation in form, say between different kinds of capitalism, we need something between this abstract theory and the concrete analysis of particular cases: that is, theories of the middle range. Thus regulation theory is a middle-range theory or analysis of capitalism which examines the different kinds of social embedding of macro-economic proesses (Aglietta, 1979; Jessop, 1990). So too are theories of industrial organization, of segmented labour markets, or of housing tenure and markets. When working with these middle-range theories it is easy to forget or take for granted the fact that they presuppose, as well as modify, the propositions of abstract political economy. Instead of asking whether planning has inherent problems we only notice particular problems of specific variants of planning. To be sure there are different forms of private property, but there are some common problems which derive from its privatism. On the other hand, in defending the big questions of political economy – for example regarding value, division of labour, exploitation, class, unpaid and paid labour, the generation of inequalities, etc., – we acknowledge that they do not tell us all we need to know about specific instances of their referents, etc.

Both radical and liberal political economy make stylized assumptions about individual motivations and behaviours. The rational-choice approach, which dominates liberal economics, makes these assumptions quite explicit. While this approach has recently been adopted by so-called 'analytical Marxism' (e.g. Cohen, 1978; Elster, 1985; Roemer, 1986 and Carling, 1992), the behavioural assumptions of more orthodox kinds of radical theory are less clear, though motivations still tend to be treated as narrowly economic in character, involving maximization of economic benefits.[3] The issue of motivation is less prominent in radical political economy because it stresses domination and downplays choice. Meanwhile sociology emphasizes – sometimes to excess – that people are also norm following and creatures of habit. As Bourdieu argues, much behaviour lies somewhere between rational choice and habit – it

---

[3] This opens radical theory to the charge that it lacks 'micro-foundations', as analytical Marxists are wont to say. However, while this is correct, rational choice theory, with its extraordinarily undersocialized view of individual behaviour, is a strange candidate to fit the bill. It merely reproduces liberal economics' impoverished view of motivation as reducing either to coercion or to free individual choice with maximization. Almost certainly, the real reason why analytical Marxists like rational-choice assumptions is that they enable them to do what they so obviously enjoy – model building and game theory based upon maximizing/optimizing behaviour.

derives from an acquired disposition or set of learned tacit assumptions and inclinations to accommodate to familiar circumstances, so that we 'choose' what is effectively chosen for us and refuse what we are refused (Bourdieu, 1984).

Virtually all the behavioural assumptions of abstract political economy of either kind could be brought into question and made into subjects of empirical inquiry. Relevant studies in sociology and social psychology have often shown that individual motivations are different to what economists assume (e.g. Burawoy, 1979; Friedland and Robertson, 1990; Lane, 1991). Does this mean that we should abandon abstract theory and try to develop a more empirically grounded political-economic theory? If the empirical investigations show the theoretical assumptions to be entirely wrong then it is hard to see what else we could do. But if actual motivations and behaviours are merely more complex than the simple representations in abstract theory, then we can claim that it remains relevant to just certain aspects of concrete behaviour. Many economic decisions have multiple motivations. We change jobs not only to get more money, but to try out a new kind of working life, to move to where we can be closer to a lover or an ageing relative. We may work hard purely for money, because our livelihood would be at risk otherwise, or in order to act honourably and to win respect. Often all these motives, and others, along with habit, influence what we do. At the same time, even if one wants to term these other motivations 'non-economic' (which might be questionable), they still have economic *effects*, whether intended or not. Conceding the existence of non-economic motives therefore does not mean we can escape from economic considerations.

We may acknowledge that political economy underestimates cultural influences within and upon economies, but this does not mean that traditional political economic theory has become redundant. We may not weigh every purchase in terms of opportunity costs, but we may be obliged to review our spending patterns when we're hard up. We may not work any harder for a higher income, but unless we have a private income we cannot ignore what work brings us in terms of the means of life. The kinds of motivation assumed in economic theory are rarely absent. Moreover the cultural influences on economic behaviour sometimes work with, rather than against, the traditionally assumed economic influences, for example encouraging us to seek more pay and higher levels of consumption. Economics is often defined as the science of the allocation of scarce resources among competing ends, political economy as the study of how societies materially reproduce themselves. The fact that in any concrete situation there is more than either of these things going on does not render those abstract sciences inapplicable, merely selective. As abstract theory, political economy does not purport to identify concrete, multiple motivations in all their richness. We can acknowledge this while rejecting reductionism and accepting that middle-range theory and empirical research have their roles too. While I have criticized *reductionist applications* of abstract theory elsewhere (Sayer, 1985; Morgan and Sayer, 1988), this book is primarily a critique of the substantive content of abstract political economy. It is not, by and large, arguing about how concrete cases are more complex than appreciated in abstract theory, and calling for more middle-range work; rather it is about whether that theory is adequate even as a partial or selective account.

Internal Relations, Disaggregative Analysis and Counterfactuals

In addition to the question of essences and abstract theory, there is a further and overlapping set of methodological problems relating to the ontological presuppositions of political-economic theory. These hinge on the issue of internal relations, and bear upon the way we analyse actual and possible economies. How far do the various elements of capitalism have autonomous identities and how far are their properties dependent on context? How far can particular elements of a given kind of economy, such as capitalism, retain their identity in other economies? Is it possible to analyse capitalism or any other system in a disaggregated way, breaking it down into particular elements and theorizing them separately? Or are economies effectively so highly integrated that, for example, in a capitalist society, everything is capitalist and cannot be considered independently of that context? These are far from merely academic questions for they have immense practical implications for possible alternative kinds of economy. If economies are not seamless wholes of internally related parts, then parts of capitalist society may operate in similar fashion in post-capitalist societies, and we can analyse such possibilities. If they are seamless wholes, then everything must change in a post-capitalist order and there are no lessons to be learned from how economic processes work now.

One angle on these questions is to be found in the conflict between analytical and dialectical Marxists. Self-styled 'analytical Marxists' advocate breaking Marxist theory down into basic propositions, analysing these one by one, and rejecting, modifying or retaining them accordingly (Carling, 1986). This flies in the face of the dialectical tradition in Marxism of treating societies as totalities and of seeing their elements as internally related. Analytical Marxists consider that anything of worth in Marxism can be explicated without resort to dialectics, which frequently serves merely as a cover for sloppy arguments (Cohen, 1978).[4]

For dialectical Marxists, the 'violence' of analytical Marxists' abstractions throws out Marx's insights into the internally or dialectically related qualities of social phenomena, and the way in which things change as they enter into new relations (Ollman, 1971; D. Sayer, 1987).[5] Bertell Ollman comes close to anti-essentialism in arguing that all objects are internally related to others and hence have no independent identity. For some objects, most famously capital itself, the internal-relations perspective is indeed illuminating, especially for making sense of Marx's own exposition of the properties of capital as it goes through different 'moments' and enters into different relations. In such cases, there seems good reason for definitions to be shifting and relational. However, if this is erected into an ontological proposition supposedly applying to all objects, it becomes absurd. If everything is internally related to everything else, then all explanations are entirely contextual and it becomes hard to see how Marxist or any other theory could be widely applicable. The relationship between my tastes in music and Marx's theory of capital, or between the stork population and the birth rate, must be internal and therefore interdependent. It is hard to imagine a more dogmatic and silly proposition.

By contrast, a critical realist ontology (Bhaskar, 1975; Sayer, 1992a) does not assume that all relations are internal but is open to the possibility of there being not only internal but external, or contingent, relations in any situation. The job of abstraction thus becomes – among other things – to establish whether they are internal or external. A realist approach is therefore more open to the possibility that features found within capitalism, such as markets or an advanced division of labour, could exist outside capitalism, and have causal powers irreducible to those of the unique features of capitalism, such as minority private ownership of the means of production.

These matters also bear upon one of the hallmarks of Marxist approaches, the concern with 'historical specificity' – nowadays taken to include geographical specificity. Other social theories dehistoricize phenomena by representing specific social forms as transhistorical essences; e.g. treating tools and machines under feudalism as 'capital',

[4] In the seventies it was common for Marxist authors to defend their ideas by saying they were *dialectical*, as if invoking this magic word, in italics of course, explained anything.
[5] Derek Sayer also adds an important criticism of those who fail to notice that Marx is addressing abstractions that in certain historically specific situations are objectified in practice as 'real' or 'concrete abstractions', and that Marxism is therefore in part a critique of those abstractions.

obscuring the vastly different social and economic significance of such means of production under feudalism and capitalism. Although its advocates appear not to realize it, a universal internal-relations ontology pushes one to the opposite and absurd extreme of assuming that everything is historically specific to the point of being unique, or entirely context dependent. But social phenomena vary in their degree of historical and geographical specificity, some being highly context dependent, others less so (Sayer, 1989). By contrast, critical realism offers support for an intermediate position in which some objects have transhistorical essences, some are specific to periods of varying duration and some are transitory.

Concrete objects therefore often combine elements which differ in their historical specificity; thus, certain features of the British state may remain while others, such as its imperial connections, change. It is a capitalist state, a patriarchal state and often a racist state, and we may object to it for these reasons, but if it were to cease to be capitalist, patriarchal and racist it could still remain a state, and retain features, some of which we might still object to, such as its power, its involvement in surveillance, or its hierarchical form. Some features of capitalist states, industry or cities may be shared by non-capitalist states, industry or cities. Consequently, if we are to sort out what is specific to capitalist instances of these, we have also to determine what is *not* unique to them. Working with an ontology of universal internal relations is likely to produce accounts which overlook this and hence fail to identify just how historically specific its objects are.

In Marx's work, one can find evidence in some places of over-extended assumptions of internal relations while in others he uses an approach more akin to that of critical realism. In critical realist vein, while he argues that certain key features of capitalism are indeed internally related, he also clearly assumes that the structures that they comprise can exist in a variety of contexts, or have some degree of context independence; that is, be merely externally or contingently related to some contexts or aspects of contexts. Although the structures he theorizes are historically specific, they are not so specific as to be unique. The fact that Marx could develop theories which can fruitfully be applied to a vast range of cases supports this claim. Thus, his theory of the labour process can inform our understanding of Toyota in Japan or an insurance company in the contemporary USA, as well as the Lancashire cotton industry in the nineteenth century. It has this generality precisely because the phenomena involved in capitalist labour processes can be reproduced in a vast range of situations, and because many of the relationships to the contexts in which they can exist are external and contingent, rather than internal and necessary. Likewise, the fundamentals of the theory of the circulation of capital apply to industries as different as agriculture and computer software. In concrete studies of such cases one would of course have to take account of the specific, and possibly unique, forms that the circulation of capital assumed, but these would not necessarily lead one to reject the abstract theory. In abstract theory we usually focus on aspects of objects or processes which can remain substantially the same under a wide range of conditions (Sayer, 1989, 1992a).[6]

Marx's method of 'successive approximations' is compatible with this for it effectively involves a series of 'thought experiments' in which certain contingent factors are held off so that a restricted set of relationships can be considered in isolation, before being examined in conjunction with others. Perhaps the clearest examples are to be found in the theory of the circulation of capital in volume 2 of *Capital* (e.g. first assuming a single universal turnover time for capital and then relaxing the assumption) (Marx, 1974). However, Marx was not entirely consistent in his anticipation of critical realist methods, for particularly with respect to the relationships between the central elements of his theory – such as class, division of labour, money, commodities and property relations – he tends to assume ubiquitous internal relations in Hegelian fashion. He thereby avoids scrutinizing his central propositions one by one in the manner of a more disaggregative approach. Perhaps in order to bolster the apparent unity of his theory, and also to avoid encouraging reformism, he generally omits to consider whether just certain parts of capitalism could be changed without changing everything else. Where he does consider alternatives, it tends to be in terms of radically different configurations, wholly different rather than partly different, for example the situation under feudal, capitalist and 'free associations' of workers in volume 1 (1976, pp. 169–71).

The ontology of universal internal relations leads one to assume that objects or processes cannot be the same under different circumstances. It simply excludes the possibility that particular processes that exist in capitalism could also exist in a society with non-capitalist social relations of production, for it assumes that they must be different, and not just superficially but fundamentally. At worst this amounts to mere tautology: everything in a capitalist society is capitalist, everything in a noncapitalist society would be different. A less circular and more interesting

---

[6] Harvey has popularized Ollman's ideas in urban studies (Harvey, 1973). He has recently modified his interpretation of the internal-relations ontology by noting that objects are affected more by proximate conditions than distant ones, but this still doesn't acknowledge context independence. This is inconsistent with his emphasis on generality and his impatience with accounts of uniqueness and difference (Harvey, 1993).

position might be that the effects of capitalism's central processes are so far reaching that everything in such a society is to some degree capitalist in character. But while this is an interesting hypothesis it is surely an exaggeration. It begs the question of whether they might actually remain much the same in some non-capitalist society (e.g. Clarke, 1982). Moreover, this position characteristically overlooks the fact that other things are pervasive too – the effects of patriarchy, rationalization, racism and sexuality, the effects of the pursuit of power for its own sake, and so on. Thus, when coupled to Marxism, the internal-relations ontology reinforces the tendency for Marxism to become a totalizing discourse, which either absorbs other phenomena as capitalist or ignores them. Further problems arise when restricted and broad or inclusive concepts of capitalism are combined in the same analysis. In another context, Delphy and Leonard (1992) attack the practice of assuming in the same breath that capitalism is both the whole social system and a determinate element in that system, so that 'all concrete forms of oppression appear to derive from capitalism while also saying capitalism is the precise mechanism of a particular form of exploitation' (p. 34). Claims regarding restricted definitions of capitalism do not necessarily license more sweeping generalizations regarding capitalism in the inclusive sense.

The existence of external as well as internal relations, and of relatively context-independent structures, justifies a more disaggregative or analytical approach than that favoured by dialectical Marxists. By this I do not necessarily mean in the style of analytical Marxism, for though this has been a stimulating influence within radical political economy, it tends not to justify the abstractions it uses as premises for its models, perhaps because of a misguided belief that saying as little as possible about premises gives the work a more 'scientific' aura (e.g. Roemer, 1986, 1988). A disaggregative approach considers whether particular elements of political-economic systems can exist only in combination with one set of other elements or whether they can also coexist with other sets; hence it assesses the validity of abstractions. If x can exist without y, then it can obviously be considered separately from y, and doing so reduces the risk of misidentifying causal responsibility between them, provided some effects are not products of the conjunction of x and y (Sayer, 1992a). Some relationships are asymmetric; thus markets can be considered separately from capitalism, but not vice versa. Where two economic systems have elements in common as well as differences (e.g. petty commodity production and capitalism), thought experiments and comparative studies are also valuable, for the situation gives more 'degrees of freedom' and shows what happens when elements enter different combinations.

A disaggregative approach therefore requires particular care over how we abstract. Abstractions may leave out that which is only contingently related to the phenomena under consideration, but if they leave out essential features which make a significant difference to the process of interest then serious misunderstanding may result. (However, not all essential features may be important or relevant; it is essential for social life that people have air to breathe but it is not relevant to most questions of political economy.) Simplification or abstraction from complexity and diversity is common in any kind of social-scientific analysis, and indeed necessary for facilitating understanding, but it is only safe if the complexity and diversity do not make a significant difference to the processes in question. In the next chapter I shall be arguing that an important feature of advanced economies is precisely their structures and mechanisms for coping with diversity and complexity; in particular, the fragmented ownership structure of capitalism and the control problems of state socialism make little sense if one abstracts from the complexity of divisions of labour. Radical political economy is flawed because it ignores rather than accounts for the practical consequences of certain kinds of complexity and diversity.

The explanations offered by radical political economy can further be assessed by considering the counterfactuals that they imply. The flip-side of any specific explanation is a counterfactual claim: if we say x was responsible for y then we imply that in the absence of x, y would not have existed or happened. If this is not the case then we know there is something wrong with the explanation. Counterfactual questions may help us see that mechanisms or conditions other than x could have been responsible for y or at least could have been jointly responsible, with x, for y. Thus, if it were claimed or assumed that the hierarchical organization of large-scale industrial production were purely a function of its capitalist social character, then asking whether non-capitalist large industry could be organized non-hierarchically would help to support Marx's view that hierarchy and supervision in such industry are also, in part, technically unavoidable. This is not to suppose that all explanations in political economy conform to this simple model of x causes (or enables or blocks) y; far from it. Often what is required is a wide-ranging narrative, in which there are many implicit or explicit counterfactuals.

In similar fashion, 'thought experiments' considering the consequences of possible changes in organization or hypothetical alternative systems can tell us much about the structure of actually existing systems. Thus, asking whether capitalism could exist without racism exposes the contingent nature of the relations between them. Alternatively, to pick an obvious example, we could establish the asymmetric internal relation between capitalism and money (money can exist independently of capitalism but not vice versa) by asking counterfactual questions (Sayer, 1992a). Given Marxism's enormous emphasis on the social relations of production it is especially useful to consider how particular outcomes would differ were these to change. In some cases what is at issue is whether capitalist social relations are relevant at all, in others whether they are the only type of social relations of production which give rise to the outcome. For example, if not only capitalism but market socialism generate macro-economic problems (Nove, 1989), then it is possible that their social relations of production are irrelevant and that the two systems share some other feature which generates those problems (e.g. market regulation).

From the point of view of a universal internal-relations ontology, it would be objected that if the social relations of production were different, then everything would be different, so it would be meaningless to speculate on the implications of changing just one component. Whatever x refers to in a counterfactual inquiry, it needs to be relatively independent of other things if it is to be the subject of such counterfactual assessments. Theoretically, this objection derives from the pivotal position given to social relations of production in Marxism and their role in distinguishing modes of production as different as feudalism and capitalism. But while feudalism and capitalism are radically different, the view that industrial societies differ no less according to their social relations of production is partly what is at issue in this book. We shall argue that there are important similarities and common features, as well as differences, among industrial societies, be they capitalist, state socialist, or market socialist.

There is, however, a more important objection to a disaggregative approach involving counterfactual inquiries. Imagine two elements which have their own effects and can exist separately, but when they do happen to conjoin, they produce 'emergent properties', irreducible to those of the consitituent elements. A natural example of this is water, which has properties irreducible to those of its constituents. A possible social example would be an enterprise or a board of trustees, which has powers irreducible to or emergent from those of its individual members. (It is often difficult to decide whether social phenomena have emergent properties since the individuals who comprise them can learn what is necessary to produce those properties, even though they can only realize them through structured interaction with others.) If x and y have emergent properties when combined, it must be remembered that their respective individual properties cannot be deduced from these. Disaggregative analyses would have to note the difference between what x can do alone and what it does in combination with y.

These methodological ideas inform later chapters' discussions of modes of coordination, ownership and comparisons of economic systems.

Critical Standpoints

A critical social science criticizes not only other theories, but actual practices or social realities, including those which inform and are informed by those theories. It starts by identifying the felt needs and sufferings of a particular group of people, and then attempts to find out what is producing those problems, be it illusions about their circumstances or material causes. In working to reduce illusions and identify causes of suffering it has an emancipatory intent and potential (Fay, 1975, 1987; Bhaskar, 1989). Radical political-economic theory could be taken as exemplifying such an approach. However, while I certainly want to endorse the idea of a critical social science, there are some major problems in resting content with this simple model.

Any criticism presupposes the possibility of a better way of life; to expose something as illusory or contradictory is to imply the possibility and desirability of a life without those illusions and contradictions. This much has been established by critical theorists such as Habermas and Apel. Yet the notion that critique implies a quest for the good is a highly abstract one. Up to a point, particular critiques do imply something a little more specific than the standpoint of a better life. The critique of capitalism's anarchic and uneven development implies a critical standpoint or contrast space of an imagined society with a rationally ordered and even process of development. The critique of class points to the desirability of a classless society. Naturally, society would be better if its illusions, conflicts and contradictions were reduced, but we naturally want to know how this could be achieved. The desirability of a life without contradictions or illusions does not make it feasible.

Critical social science does not merely identify illusions, irrationality or contradictions but attempts to provide explanations of their sources, locating the 'unwanted determinations' of behaviour, as Bhaskar (1989) puts it. It would be strange, to say the least, if an analysis of the causes of problems such as hunger and exploitation were unable to indicate anything about alternatives which would eliminate them. If a critical theory cannot begin to indicate how to eliminate problems we must inevitably be suspicious of its claims to have identified their causes. If the alternative implied by a critical standpoint is not feasible, then any critique made from that standpoint is thereby seriously weakened. Not to put too fine a point on it, the critique of, say, capitalism's anarchic and uneven development would lose much of its force if all advanced economies were necessarily anarchic and uneven in their development, though one could still criticize advanced economies – not just capitalist ones – from the very different standpoint of a 'deep ecology', calling for a return to small-scale, more primitive economies (Dobson, 1990).

We need to know enough about the critical standpoint and the implied alternative to be able to judge first whether it really is feasible and desirable. Since knowledge is 'situated' and bears the mark of its author's social position, this includes assessing whose standpoint it is made from. Does it privilege the position of a particular group (e.g. male workers, advanced countries)? Does it imply a society without difference? If it suggests greater equality on whose terms is equality to be defined?[7] We have also to ask whether remedying one set of problems would generate others (it usually does), and whether these would be worse than the original problems. This is rarely considered in radical political economy, the usual implicit assumption being that all bad things go together in capitalism and all good things under socialism/communism. Yet it is possible that some of the 'contradictions' involve dilemmas which can't be eliminated along with capitalism. Evaluations in terms of desirability therefore need to be cross-checked with assessments of feasibility, and optimistic assumptions of inevitable improvement suspended.

There are two kinds of feasibility which might be considered:

1 whether a certain desired end-state or goal can be realized – for example, how people can be politically mobilized to make it happen; and
2 whether, assuming enough people are willing to try to make it happen, the goal or end-state is feasible in itself, e.g. could one have an advanced economy without money?

It is usually only the first of these questions that radicals address, the standard response to utopian discussions being not 'would it work?' but 'yes but how are you going to get from here to there?' But while many might think it idle to ignore (1), it is surprising how little attention is given to (2), as if the journey mattered more than the destination. I fully accept that I am not offering suggestions on (1) in this book, and only ideas pertinent to (2): but then I don't see how large-scale political mobilization can precede a well-worked out conception of a feasible alternative.

All this is not to argue that the absence of plausible superior alternatives should render critical social science silent in the face of the ills of capitalism or whatever. The lack of such alternatives might blunt

---

[7] Some Marxists continue to ignore these problems, e.g. Callinicos (1991).

the political force of its critiques but it would still be important to have a critical social science which identified the causes of social ills, even though, for the time being, we could not see a way of eliminating the causes without generating greater problems. However, I don't think the situation is this desparate – there are at least some promising alternatives.

Many Marxists resolutely refuse to discuss the nature of their preferred alternative society, arguing that to produce 'blueprints' would be to pre-empt the course of class struggle. But what we are calling for here is not blueprints but merely attempts to think through the likely tendencies or mechanisms of different forms of political-economic organization. For example: Can highly complex economies be subject to detailed *ex ante* control by participatory democracy? Could a change in property relations eliminate externalities? Could markets be regulated so as to avoid regressive effects? All too often the blueprints argument merely serves to disguise the fact that those who advance it haven't got a clue about feasible alternatives. Is it not strange that a critical social science which is so exacting in its explanations and critiques of what is, should be so lacking regarding what could or should be? As Nove asks 'what is one to make of a "scientific socialism" that seeks wholly to escape rational discussion as to its own feasibility and working rules?' (Nove, 1983, p. 14). To be sure, we cannot anticipate the future but we can make some judgements about what is or is not feasible and desirable. Marxists are fond of quoting Marx on the difference between the labours of bees and human architects (Marx, 1976, p. 284) (only humans imagine what they are going to build before they construct it), and yet when it comes to alternatives Marxists would appear to prefer to be bees. Worse still is the practice of sliding between an idealistic and vacuous definition of socialism as a society free of contradictions and injustices, and a substantive definition of a system which could actually exist. This ensures that the failings of any actual, attempted socialist system count for nothing, because, naturally, they aren't considered to be truly socialist.

As Steele (1992) points out, arguments about the merits of utopianism frequently get confused by two different concepts of utopia. If utopias are treated as being infeasible by definition then that prejudges any questions about their merits; merely to mention the word utopian is then to slam the door on rational assessment. If, on the other hand, we leave it open as to whether utopias can be feasible, we avoid excluding possibilities and enable a more rational assessment to be made, and can challenge their proponents to make them feasible. Considering utopias is therefore compatible with 'science' since it is consistent with asking counterfactual questions, conducting thought experiments and scrutinizing critical standpoints. Indeed, as Steele argues, Marxism would be more scientific if it were more utopian (Steele, 1992, pp. 374–5).

Critical standpoints and alternatives don't have to drop from the skies or be vacuously vague. They can come from what Wellmer terms a 'draft meaning' – a particular area of existing progressive practice, which provides a model which we would like to generalize to the whole of society (Wellmer, 1972). Thus the equal treatment of people as voters might be taken as a draft meaning for the extension of democracy to industry.[8] The general problem with draft meanings is that it is tempting to assume that just because a particular practice or institution is progressive, it must still be progressive when extended to other situations. In practice it may be desirable or feasible only as a 'niche' phenomenon rather than an embryonic version of something potentially large.[9] Hence regarding the extension of democratic control, it is necessary to consider whether the costs of participation to individuals are likely to outweigh the benefits if they have to vote on large numbers of complex issues which are currently dealt with through delegation or division of labour. In considering draft meanings, we therefore have to assess the possibility that the generalization or extension of a currently restricted phenomenon may be either infeasible or else generate side effects that are as bad or worse than the problems they are supposed to resolve.

Evaluations of whole systems need to be exactly that, not judgements of the whole based upon just a part of the system. It is not enough, for example, to argue that if the capital–labour relation can be shown to be unjust, it therefore follows that capitalism must be overthrown (see e.g. Carling's case against capitalism, 1992, pp. 145–6). Capitalism in the inclusive sense consists of more than the capital–labour relation and the ignored features may produce effects which offset the ills of exploitation within the relations of production. It also has to be demonstrated that there is a superior alternative. All too often critical theorists ignore these points and judge and condemn whole (concrete) systems on the basis of abstract analysis of just one of their parts, while invoking an unspecified, imaginary society as automatically superior. As Buchanan notes, Marx and later Marxists 'simply assume not only that a highly productive alternative to the market is feasible, but that a high level of productivity and a system for efficiently distributing what is produced can be achieved by a system of democratic decision-making' (1985, p. 29). To

---

[8] However, critical theorists failed to appreciate how draft meanings could take regressive, as well as progressive forms. The drift of politics over the last fifteen to twenty years, from the rise of the new right to the resurgence of nationalism and 'ethnic cleansing', provides all too many examples of the diffusion of regressive draft meanings.

[9] This is consistent with Michael Walzer's concept of different 'spheres of justice' in which different moral principles are appropriate. Thus what is appropriate for a club might not be appropriate for a neighbourhood or country (Walzer, 1985).

be useful, evaluations or critiques need to involve 'intersystemic comparisons', in which the systems are evaluated in equal depth, involving comparisons of balance sheets of strengths and weaknesses instead of isolated points of condemnation (e.g. Buchanan, 1985; Putterman, 1990). Evaluations of economic systems have to be balanced, not in the sense of having to arrive at a neutral conclusion, but in terms of considering *pros* and *cons* of different forms of organization. At the end of the twentieth century it should be glaringly obvious both that social change is inherently contradictory and 'dilemmatic' (Billig, 1988), and that while those contradictions and dilemmas continually mutate there is no sign of any reduction in their number or severity. This should not induce defeatism, just a greater regard for the need to consider intersystem comparisons rather than isolated points. This might seem trite, were it not for the tendency of political-economic literature to be polarized into apologetics on the one hand and critiques which acknowledge only the bad.

To summarize so far: Assessing radical political economic theories as a critique of capitalism involves asking:

A: whether capitalism necessarily causes the problems in question;
B: whether capitalism is unique in this respect or whether certain other social systems would generate the same problems;
C: whether any conceivable alternative could avoid them;
D: whether resolving those problems would create other, worse problems.

Simply identifying problems and their causes within capitalism (A) is not enough. Clearly, if alternative systems produced similar problems the ills would still be repellent, but it would not make sense to single out capitalism for attack.

Critiques draw upon a variety of sources and Marxism is no exception. As Allen Wood has pointed out, in Marx, these include unexceptional values implied by his condemnation of the way in which capitalism causes 'avoidable death, waste, hunger, mental and physical exhaustion, monotony and loneliness' (Wood, in Buchanan, 1982, p. 26). But it is the standpoints that we can glean from the limited and fragmented writings on socialism and communism that distinguish Marxism from other critiques.

Traditional Marxism implies the coming into being of a world-historical or collective subject through the overthrow of class divisions and the transcendence of division of labour (see e.g. Mandel, 1976). This view has taken a hammering in the last decade, and deservedly so. It has been attacked on several accounts. The following criticisms are a mixture of liberal and postmodernist points:

1 It is infeasible: modern societies are too complex ever to be organized as a single, collective historical subject. Even the elimination of class would leave societies divided along many other lines. Even in the absence of direct antagonisms between groups, one cannot expect a consensual community to form at a national scale, let alone a global scale;
2 on moral grounds: the goal of a collective subject is inherently authoritarian and conservative because it cannot permit difference, or value pluralism;
3 on epistemological grounds: knowledge is too complex, diverse, disputed and uncertain for it ever to be possible for a collective subject to plan social development consensually and rationally. As we show in the next chapter, associated with the division of labour is a division of knowledge which is too complex and diverse for any single agency – for example, socialist planners – or indeed any electorate – to cope with.

In Marxist political economy, the draft meanings come from an uncomfortable combination of two discrepant sources:

First the model of unalienated, apparently pre-industrial social individuals, freely producing for known others according to socially agreed goals and norms and without the mediation of markets, and moving between a diverse range of jobs. This is most closely associated with the (in)famous pre-industrial parable of the person who is a the hunter, cattleherd and critic, all in one day, of *The German Ideology* (Marx and Engels, 1974), but is evident in the later Marx too (e.g. *Capital* vol. 1, 1976, pp. 171 and 618).

Second a society with highly developed forces of production, in which the model of planned, rationally organized cooperative labour in the factory, in which workers submit to a technically determined discipline, is stripped of its 'capitalistic shell' and scaled up to the level of society as a whole.

Marx and Engels never adequately reconciled these standpoints: as we shall see in the next chapter, they cannot be reconciled. Although Marx was scathing about those who wanted to return to pre-industrial kinds of economic organization, he failed to break completely with the critical standpoint of a pre-industrial social organization. The picture of a society without alienation is one of a minimal division of labour, as implied in Marx and Engels' hunter/fisher fable, for it is not merely class but division of labour which causes alienation (Marx and Engels, 1974). As Habermas puts it: 'The socialism about whose normative content Marx refused to speak is Janus-faced, looking back to an idealized past as much as it looks forward to a future dominated by industrial labour' (Habermas, 1991, p. 38).[10]

[10] Habermas adds: 'The idea of a free association of producers has always been loaded with nostaligic images of the types of community – the family, the neighbourhood and

There is also a more rationalist interpretation of the communitarian element in Marx's thought. This sees community not as bound together by traditional norms but as a future society achieved through the unleashing of men's and women's capabilities for rationally planning their society and environment, rendering it transparent to all. Through the development of their rational powers, they would realize their capability for 'altruistic, directly cooperative motivation, not opposing their interests to those of "the good of the larger society"' (Buchanan, 1985, p. 107). All that blocks this, apparently, is 'ignorance and defective social conditions' (ibid. p. 108).

Alienation in Marx's sense is unavoidable in an advanced industrial society, whatever the social relations of production, since economic processes necessarily become ever more specialized, more mediated or 'roundabout', more routinized and rationalized, as a condition of the possibility of high levels of material development. The critique of alienation has now lost much of its force, not least because of its questionable view of human nature and the many compensating benefits of living in an advanced economy (see Kymlicka, 1990, pp. 186–92). Whatever residual appeal this standpoint has today lies in its resonance with a longing for community, which remains strong in contemporary social theory and everyday life. As we shall see shortly, it is a hopeless ideal for societies with advanced economies.

Conclusions

I have argued that notwithstanding the legitimate rise of middle-range theory relating to political economy, and the anti-essentialist critique, there is still a place for abstract theory, provided it is not used in a reductionist fashion. Secondly I have made a case for a 'disaggregative' approach to political economic theory, against the resistance to such a strategy offered by dialectical or internal-relations theorists who favour a totalizing approach. We need to ask counterfactual questions and conduct thought experiments about possible alternative systems, refining ideas about the specifically capitalist nature of phenomena such as the

---

the guild – to be found in the world of peasants and craftsmen that, with the violent onset of a competitive society, was just beginning to break down, and whose disappearance was experienced as a loss. The idea of the preservation of these eroded communities has been connected with "socialism" ever since these earliest days' (Habermas, 1991, p. 38). Similarly, Kamenka characterized Marx's theory as combining 'technical scientism with anarchistic peasant communism' (Holton and Turner, 1989, p. 167). See also Miller, 1989, p. 29).

state, cities or industry by asking what is not specifically capitalist about them. Thirdly, I have argued that as a critical social science, radical political economy needs to examine its critical standpoints and the alternatives implied therein to check whether they are feasible and desirable. Since any complex economic system is likely to have inherent problems, evaluations need to be based on intersystem comparisons rather than on isolated criticisms of a single system. Marx's own theory is seriously lacking on all these counts and includes two fundamentally inconsistent critical standpoints. Frequently, explorations of counterfactuals or thought experiments will throw up dilemmas where goods and bads are interdependent, such that eliminating the bads also eliminates the goods. One of the naïve features of theories of early critical social science was the implicit assumption that such dilemmas would always be resolvable (Fay, 1975, 1987). On the contrary, we should approach critical social science in the expectation that social life is dilemmatic. Opponents of critical theory have argued that even rational, well-informed people may not always be able to reach a consensus on what is wrong and what should be done, because there may be no clear superior among the competing views of society and moral philosophies. But even if there were such a consensus on viewpoints, the 'objective' dilemmas of social life, with their complex and often inseparable combinations of good and bad, would still make consensus on diagnoses and prognoses unlikely.

Finally, I want briefly to draw out some of the links between the points and proposals made in this chapter.

1. Critical social science is often defined as starting from a problem affecting people, such as racism or exploitation. What is defined as a problem depends on the critical standpoints of lay persons and researchers. These standpoints involve moral and political judgements. Such evaluations are not disconnected from positive (descriptive and explanatory) accounts of practical situations but actually make assumptions about their nature. If I say that the capitalist's expropriation of the workers' product is wrong, I am not merely expressing a feeling but making assumptions about what is going on in that situation. In shorthand, values may presuppose facts, or normative judgements can presuppose positive (descriptive, explanatory) statements.

2. Critical social science offers explanations of the causes of the 'target' problems, and its claims regarding those causes can usefully be assessed by means of counterfactual questions and thought experiments.

3. Counterfactual enquiries and thought experiments not only check causal explanations but help assessments of the feasibility and effects of alternative, perhaps as yet untried systems. In so doing they assist in their moral and political evaluation by giving a clearer picture of their effects. For instance, imagining the likely nature and effects of markets without capitalist social

relations of production is useful not only for assessing radical political economy's explanations of how markets work under capitalism but for considering whether it would be desirable to retain markets outside capitalism.

4. Examining the feasibility and desirability of alternatives can reveal additional, hitherto unnoticed problems in them. Thus, we may conclude that fragmentation or individualism in capitalist society is indeed a serious problem, but if its elimination generates worse problems, then we can say that the initial critique of those problems is overridden – though not undermined – by arguments concerning countervailing processes.[11] This reaffirms our earlier point that whole systems can't be persuasively condemned merely on the basis of one or two sources of problems and without comparisons with the *pros* and *cons* of alternative systems.

5. Positive claims do not logically entail particular moral judgements – it would not be illogical to agree that workers produced surplus value but disagree that it was wrong. But positive claims can *influence* moral judgements. If y is already deemed to be bad, and x is claimed to be its cause, then the negative judgement of y is liable to be enlarged to include x. For example, we might already consider lung cancer to be bad, but once we know that it is caused by smoking, then our evaluation of smoking is likely to become similarly negative. Having initially thought that poverty was bad and markets good, our evaluation of the latter might change if it were found that markets were responsible for poverty.

6. Looking at how the relationship between evaluations and explanations works from the other direction, predispositions as to what is good and bad tend to encourage us to favour or disfavour explanations accordingly. Critical standpoints and moral and political values therefore orient research, consciously or unconsciously. A hatred of class might predispose us to blame a wide variety of social ills on it. A belief in community might predispose us to say that lack of community is a cause of social ills. In either case, the causal explanations of social ills are not justified or logically entailed by the values, but they are attractive to holders of those values because by linking class or lack of community to social ills they reinforce the conviction that class is bad and community good. On the other hand, value positions do not necessarily obstruct objectivity; in some cases they can assist it. Researchers committed to feminism are likely to identify aspects of society which others simply miss, so that they increase the objectivity of social theory. But again there are no guarantees here: value positions do not logically entail sound explanations, and may indeed encourage wishful thinking.

7. I have argued that our critical standpoints and their implicit alternatives need more attention. Later I shall argue further that political economists need to engage with normative theory, particularly political theory, so as to clarify

---

[11] See Taylor (1967) for a fuller discussion of undermining and overriding critiques and the links between science and values.

what they consider to be good or bad in social arrangements. However, lest I be misunderstood here, I am not suggesting that political economists should decide on a particular alternative or decide whose side they are on (the working class, women, the Third World, or whatever), and then proceed on that basis, explaining and evaluating everything accordingly. It is not a matter of choosing a particular critical standpoint, nailing our colours to the mast (. . . as a socialist, I believe . . .'), and then letting rip with a critique. Such assurance about moral and political values and alternatives is wholly unwarranted. Rather, I am arguing for an examination of a *variety* of alternatives, one which *problematizes* assumptions about what is good or worthy, rather than taking them for granted or as a matter of personal affiliation. I shall therefore not offer a single favoured alternative and use it as a standpoint for a critical analysis of existing economic systems, but explore several alternatives or standpoints.

3

Division of Labour and Economic Power: A Reconceptualization

Perhaps the most extraordinary and impressive feature of modern economies compared to old is their remarkable division of labour, spanning and integrating people and economies across the world. Sociologists, particularly Durkheim, have attached great significance to it, though without providing many concepts useful for economic analysis. In radical political economy the treatment of division of labour is cursory to say the least, forming little more than a backcloth in front of which principal players such as class, gender and race move (Sayer and Walker, 1992). While there are many angles from which one could criticize radical political economy, including the treatment of class and gender, it is my view that the conceptualization of division of labour is deeply problematic too, though it has attracted little attention. I shall argue that, ironically, radical political economy has an insufficiently materialist conception of division of labour which leads it to underestimate the intractability of advanced economies of all kinds. It is rarely appreciated that the distribution of power, including class power, in an advanced economy is heavily influenced by the division of labour and the division of knowledge associated with it. As long as radical political economy fails to take note of this, its critiques of capitalism are likely to be made from incoherent, infeasible and in some ways undesirable standpoints.

To develop this critique we must examine the conceptualization of division of labour in considerable detail. In the first section I present a 'first cut' at this, outlining major kinds of division of labour, and distinguishing them from another form of social division with which they are often confused – class. I then discuss the significance of the complex divisions of labour of advanced economies, in particular their intractability and their effect on the distribution of power. In the ensuing sections, we look at three different lines of theorizing which converge to present a more materialist understanding of divisions of labour: the first concerns the implications of economies and diseconomies of scale and scope for economic power; the second unpacks the different criteria implicit in Marx's distinction between technical and social divisions of labour; and the third looks at the possible contribution of Hayek's distinction between economy and catallaxy. I then examine some of the implications of these issues for radical political economy's critical standpoints.

Division of Labour: Provisional Definitions

Division of labour refers to situations in which individuals undertake different specialized types of work and hence become dependent on one another. There are many kinds of division of labour: between individuals in households with respect to domestic work, between workers in enterprises and institutions within the formal economy, between mental and manual or executive and subaltern work; there are also divisions of labour based on gender, race or age, and so on. Although they inevitably imply exchange of some kind, albeit not necessarily involving money, there is no reason why the relationships should be equal and free of domination; on the contrary, divisions of labour can be constituted by relations of domination, as in the gender division. Of all these types I shall pay most attention to the divisions of labour within and between enterprises, as it is here that I believe radical theory to be most deficient.

As a source of 'first cut' definitions we can generally follow Marx, though later we shall need to refine them considerably. Marx made a major contribution by identifying the differences between divisions of labour within and between enterprises. The division within a capitalist firm, called by Marx the detail or manufacturing division and now commonly known as the technical division of labour, is planned and controlled by the owner. Each individual worker makes a partial contribution to the production and sale of the commodity rather than producing the whole product or service. In its horizontal dimension, the technical division involves workers carrying out different tasks, but there is usually also a vertical dimension of supervision and management of those tasks. Following Marx, the division of labour between enterprises is generally known as the social division of labour. Here, under capitalism, the workers relate to one another as producers of different products, working for separate capitalists.[1] Moreover,

---

[1] Note that contrary to a common misconception (e.g. Devine, 1988), the technical–social distinction does *not* correspond to the difference between differentiation of tasks and social hierarchy.

While within the workshop, the iron law of proportionality subjects definite numbers of workmen to definite functions, in the society outside the workshop, chance and caprice have full play in distributing the producers and their means of production among the various branches of industry.

(Marx, 1976, p. 476).

This gives rise to the widely-noted contrast between the 'despotism' of the capitalist's *a priori* control of the technical division, and the 'anarchy' of the *a posteriori* regulation of the social division.

The Marxist terminology is far from ideal. There is nothing asocial about the technical division: it is responsive to social differences such as gender divisions, as well as to technical influences such as the numbers of people needed for the different stages of making a product. Moreover, although the social division can exist between firms involved in similar work, it also divides sectors which differ in their technical character, such as mining and food processing. A further source of uncertainty in Marxist discourse, this time involving more than terminological problems, concerns whether division of labour belongs with the social relations of production or the forces of production. Any division of labour involves production and social relations between producers, and so might be included under social relations of production. On the other hand, the development of cooperation and specialization, i.e. 'social labour', is a major component of the development of the forces of production. Not surprisingly, Marx sometimes referred to the effects of division and cooperation as the 'social productive forces of labour' in *Capital* (Marx, 1976, p. 1024; D. Sayer, 1987, ch. 2). While it is important to note this dual quality of division of labour, as part of both the relations and forces of production, we shall follow the custom of reserving the term 'social relations of production' for those between owners and non-owners of the means of production, i.e. class relations in Marxist terms.

Notwithstanding these difficulties, the technical/social distinction has proved to be enormously fertile for radical analyses of capitalism. In particular, the contrast between the planned and despotically controlled nature of the technical division and the *ex post*, 'anarchic' market regulation of the social division of labour illuminates capitalism's extraordinary and perhaps contradictory combination of spheres of rational organization and spheres of anarchy, a combination which structures both the world economy and the texture of everyday life. It therefore has much to contribute to the understanding of political economy in general and uneven development in particular.

In most political-economic theory, divisions of labour are treated as passive products of the machinations of class and patriarchy (Sayer and Walker, 1992). But divisions of labour play an active role in their own right. As some occupations and industries are more strategically placed than others in relation to the functioning of the economy, divisions of labour give considerable scope for inequalities to arise. Their niches create scope for the formation of different interest groups and provide them with a basis for action and for social closure against others (Parkin, 1979). Thus we find conflicts across the social division of labour such as those between US steel workers and Japanese car workers, or between nuclear power workers and workers in other energy industries. Technical divisions of labour create scope for conflicts and inequalities within firms, between the marketing department and engineering department, between maintenance workers and operatives, and so on. As in any structuration process, these divisions of labour are partly the product, as well as a medium and cause, of power differences. With varying degrees of success, the occupants of particular parts of a division of labour try to organize and influence its further development and hence their economic power. Inequalities, rivalries and tensions are the norm.

Divisions of labour in the formal economy tend to 'embed' and be shaped by wider divisions in society. Most obviously, different kinds of work become gendered in particular ways. One of the means by which patriarchy is reproduced is by men taking advantage of the uneven opportunities provided by division of labour for gaining economic power; for example, the tensions between doctors and nurses, or between maintenance workers (generally male) and operatives (often female) are strongly gendered. Similarly, racism operates to different degrees and in different ways across the division of labour, with ethnic minorities being excluded from some occupations and ghettoized in others. However, it is important to note that there would be inequalities and power differences across the division of labour even in the absence of gender and race divisions.

Divisions of labour presuppose, as a condition of their existence, modes of coordination or integration which can link up the various specialized activities in time and space. Much has been written in radical political economy on the macro-economic coordination of capitalist economies, in terms of modes of regulation (Aglietta, 1979; Boyer, 1990; Lipietz, 1987; Jessop, 1990). My concern is with the micro-economic modes of coordination which regulate material and information flows, control rates of exchange, and provide signals, rules and incentives to actors, so that specific products meet up with specific users with specific needs. At this point I shall merely outline some of their key features, leaving a fuller analysis for chapter 4.

Micro-economic coordination is not simply a matter of producing the right proportions of broad categories such as pharmaceuticals, electronic goods, construction materials, vehicles, holidays, plastics, insurance, machinery or 'miscellaneous goods'. All economies require users in particular places with highly specific needs at particular times to be supplied with those specific products. A sick person does not need just 'medicine' but a particular kind of medicine, and no other. If something goes wrong with a component in a computer, only one kind out of thousands of electronic components will do to replace it. There are thousands of different products used by the construction industry, but only a particular selection from that vast range of products was used to build the house I am sitting in. In political economy and social research we are accustomed to abstracting from this specificity, but our everyday lives depend on micro-economic coordination at this level of detail.

The best-known modes of coordination are markets and central planning, but others exist too. Divisions of labour may also be coordinated by various combinations of custom, authority, coercion, negotiation or democratic process (see Lindblom, 1977; Dryzek, 1987; Sayer and Walker, 1992). These latter modes are often ignored by economists, who treat the market and planning modes as self-sufficient. However, both of these presuppose trust and acceptance of a certain moral order (custom, authority), rely on the coercive threat of the state, and involve considerable negotiation. The converse does not necessarily apply, however: custom, authority, coercion, negotiation and democratic process can coordinate some divisions of labour without markets or central planning. These commonly overlooked modes can both support and 'embed' planning and market exchange, and exist independently of them.

Modes of coordination not only join up producers and users but influence the allocation of scarce resources (including labour power) among competing ends, thereby affecting who can produce or consume what. This is not simply a matter of substantive rationality in choosing among ultimate ends, but of deciding between an often vast number of possible means of meeting those ends. Means and ends are not entirely separable here, for the extent to which ends can be met depends on how parsimoniously resources are used in meeting them. The level of development of a society therefore depends in part on this allocational efficiency. Socialists are often tempted either to dismiss this issue as 'merely technical' — as if the problems of allocation in a complex modern economy could easily be solved! — or to treat it under socialism as a matter purely for political decision — as if allocational efficiency should always be sacrificed for political goals, even though allocational inefficiency might render those goals unattainable.

Any mode of coordination requires a fairly clearly established definition of rights and responsibilities over the property being produced, used, exchanged or transferred. It must be clear, if chaos is not to result, who is allowed to do what, with what, so that the division of labour is unambiguous. In markets these matters can be contested through competition, but at all times ownership must be clearly defined. Central planning requires a bureaucratic system in which duties and powers are clearly specified.[2] Networks are a more flexible mode which allow the precise division of labour and responsibilities to be negotiated explicitly. In all cases we can expect the establishment of divisions of labour to be contentious and disputed to some extent. No mode of coordination is perfect or ever a neutral medium for the transmission of power; they can disperse or concentrate power, and amplify or reduce inequalities.

The remarkable increase in division of labour in the last two centuries has been enabled by the development of modes of coordination, particularly markets and bureaucratic control.[3] (This relationship can be reversed, however; if severe coordination problems occur, then the degree of division of labour must fall. Thus, under state socialism, problems of getting supplies and parts forced many enterprises to make in-house parts which they were supposed to get from other enterprises.) This does not mean that the effects of division of labour are wholly reducible to those of the modes of coordination present, as it may be possible to substitute one mode for another within roughly the same division of labour. This implies that comparing economic systems purely on the basis of mode of coordination, as in the popular plan-versus-market framework, is unsatisfactory. Firstly it has the effect of underestimating the influence and relative independence of other dimensions, such as the form of ownership. As we shall see, ownership forms do not correspond neatly to a plan–market distinction. Secondly, it tends to lead one to underestimate the problems which divisions of labour present under all modes of coordination. For these reasons we regard divisions of labour as more fundamental than particular modes of coordination like markets and planning.

Class

I shall use the term 'class', *not* as a way of classifying the whole population, but in a very restricted sense, to refer to groups with shared positions with respect to ownership or non-ownership of means of

---

[2] Issues of ownership and control are taken up more fully in chapter 6.
[3] Economists often ignore bureaucratic control. Weber, Marx and economic historians such as Alfred Chandler provide good correctives.

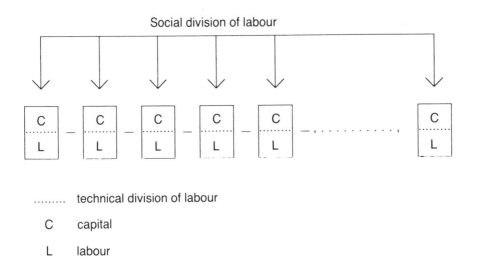

Social division of labour

.... technical division of labour

C   capital

L   labour

Figure 3.1   Class and the social division of labour

production, such as wage labourers and capitalists, or the self-employed.[4] Ownership of means of production is a source of power in its own right, for it generally allows owners to use and dispose of their property – a kind of property to which others may need access in order to live – as they see fit. Where ownership of much of the means of production in a society is concentrated in the hands of a minority, the livelihoods of the dependent, propertyless class are under the control of the owners. Under capitalism, other things being equal, owners can hire and fire workers and, to some degree, use this as a threat to make them work as they wish. The causal powers of ownership are not absolute, however: sometimes they are limited by state regulation (e.g. regarding employment legislation); sometimes they are rendered worthless, as in the ownership of obsolete means of production; and sometimes they can be overridden by other forces, such as powerful occupational groupings or organized labour.

Figure 3.1 provides a simplified representation of the relationship between class and the technical and social divisions of labour. In the horizontal dimension we find the thousands or millions of separate

[4] A fuller discussion of this concept of class can be found in Sayer and Walker, 1992, ch. 1.

enterprises or capitals constituting the social division of labour, producing different products and linked in most cases by market relationships,
and within each enterprise there is a technical division of labour. In the
vertical dimension, within each unit or enterprise, are class relations
based in capitalism around minority private ownership of the means of
production. This diagram counters the view implicit in Marxism that
there is a natural unity to the working class. As Meszaros notes, much
of what Marx wrote about the fragmented nature of the peasantry also
applies to the working class under capitalism (Meszaros, 1987). Indeed
wage labour can only exist on any scale where this fragmentation by the
social division of labour has taken place.[5]

A long line of theorists have produced more disaggregated and less
restrictive accounts of class than mine, but they have often done so by
conflating class with other sources of social division, especially division
of labour and the skill differences therein. If our concept of class seems
unusually narrow, it is because we have unburdened it of these other
sources of inequalities with which it is commonly confused. We do not
imagine that class has a privileged place in structuring the whole of
society, so that everyone could be classified according to it.[6] Nor do we
attempt to make it subsume and explain all forms of domination and
inequality within the sphere of production in the formal economy.
Divisions of labour, patriarchy and racism are also sources of inequality
in their own right, within the sphere of production as much as outside.

It has often been noted that labour power is itself a kind of means of
production; Marx himself acknowledged this in referring to the workforce seen from the standpoint of capitalist production as 'variable
capital'. From the point of view of 'human capital theory', which stresses
how the skills and knowledge of workers differ and how their power
varies accordingly, it is tempting to replace or supplement Marxian
versions of class theory with one that proposes a range of classes
corresponding to different levels of human capital. While the skills and
know-how of workers certainly do differ considerably and affect the
power they have, they are rarely sufficient on their own to enable
workers to become capitalists and employ others. To do so they need
more conventional material means of production too. (It is striking that
those kinds of work in which 'human capital' is overwhelmingly the
main kind of means of production, as in design consultancies or

[5] In the absence of the social division of labour, wage labour could only exist as in
domestic service, that is, in an unspecialized form where its products are immediately
consumed by employers rather than being sold to others.

[6] Thus, those, such as children, who do not work, do not have a class position in their
own right, though they could be said to have 'associate membership' of the class(es) of
their parent(s) (Sayer and Walker, 1992, ch. 1).

hairdressers, capitalist social relations of production do not develop on a large scale.) In most cases, the power which individual workers derive from special skills and knowledge, such as knowledge of accounting, can strengthen their class power, i.e. the power that they have as wage labourers, and it gives them access to the best positions within the technical divisions of labour within organizations, be they capitalist or otherwise, but without changing their class position in our sense. (Employee managers have an ambivalent position because they take on some of the functions of capital without being owners, but because they are not owners they can still be sacked.)

It therefore makes sense to distinguish class and class power, depending on ownership or non-ownership of the means of production in the conventional sense, from power deriving from skill and knowledge. The power of the so-called 'service class' (Lash and Urry, 1987) is real enough, but in our terms it is not a form of class power. To infer then that professional employees have the same class position (on our definition) as employee cleaners is not to attempt to conceal the differences in their power so as to create a spurious picture of a homogenous and unified workforce. The differences in their power are huge and highly significant, but they involve non-class sources of power – skill, knowledge, division of labour, as well as gender and ethnicity. Indeed we can best affirm the importance of non-class differences by not confusing them with class differences, and vice versa. Any one individual has several sources of power or lack of power, to do with class position, division of labour position, skills and knowledge, gender, race, age, beauty and so on, any of which can be enhanced by organizing with, and/or against, others. The multiple sources of power cannot sensibily be subsumed under a single one of these headings, though this is exactly what much of 'class theory' attempts to do. Those who resist our idea of treating professional employees and employee cleaners as in the same class probably do so because they cannot drop the habit of using an overburdened concept of class which attempts to cover all differences in power, income and life chances.

That class and division of labour characteristics are separate and relatively autonomous can be seen by considering the situation in non-capitalist enterprises, for these represent practical demonstrations of the relative autonomy of divisions of labour from class. Thus a cooperative could have managers and workers, manual and non-manual workers, skilled and unskilled workers, it could have variations in the amount of autonomy and discretion over work, and yet every individual could be equal in terms of ownership. To be sure, a cooperative might very well try to limit these division of labour differences, precisely to stop them overriding its members' equality in terms of ownership or class. Although power deriving from class and division of labour can interact, they are still distinct. Owner-managers have an additional source of power compared to employee managers. And, other things being equal, a cleaner who is a co-owner in the enterprise in which he or she works has more power than one who is an employee. For these reasons, differences arising from division of labour, such as that between skilled and unskilled workers, should not simply be subsumed under class. Divisions of labour really do divide labour, though not in the same way that class does. Maintenance engineers cannot hire and fire machine operators, but they can make their lives difficult. Highly productive workers in firm A may be partly responsible for less productive workers in competing firm B losing their jobs, though it is the owners of firm B who actually have the power to fire them.

There is a second respect in which our concept of class might seem unusually narrow, for unlike many class theories, it does not claim to enable us to predict individuals' life chances, standard of living, life-style, etc., from their position at work. If we are to understand the relationships between class and division of labour and between these and other sources of social division such as patriarchy and racism, it is vital to break with the assumption that a theory of class should not only single-handedly explain peoples' power within production but should also explain the differences in their incomes, life chances, etc. In many cases this assumption seems to arise from a curious belief that analytical, social-scientific concepts of class have to be ratified by reference to everyday uses of the term class, which have more to do with life chances and life-styles. While it is important to explore these everyday uses,[7] they are notoriously diverse and inconsistent, so we should not expect them to correspond to production-based concepts of class. In other words, our concept of class is abstract, not concrete; it is not intended to serve as a kind of stratification theory comparable with those of many sociological class analyses.

These two kinds of over-extension of the concept of class – from being one among several sources of power within production to being treated as the sole source of power; and from the sphere of production to life chances – are linked. If one wants to explain life chances by reference to individuals' place in production it obviously helps to load many other sources of inequality in the workplace under the heading of class. The cost of doing so, however, is conceptual confusion in which the effects of differences in power in different parts of divisions of labour are lumped together with relations of domination deriving from ownership of means of production.

---

[7] See, for example, Bourdieu, 1984.

There are therefore two main reasons for rejecting the deeply engrained, heroically broad concept of class. One is its blindness to gender and other non-class forms of social division and domination which produce, in their own right, major differences in life chances. To be sure class in our narrow sense has some bearing upon life chances, but so do several other different social structures, and therefore trying to explain them purely by reference to class produces major errors and absurdities (Walby, 1986, 1990). The other problem with this broad concept of class is its conflation and confusion of class (and sometimes non-class relations of domination) with division of labour.

I now want to argue that although class, gender and division of labour interact in complex ways they are actually separable structures, capable of independent existence. It is clear that patriarchy, racism, ageism, homophobia and so on, can exist independently of capitalism, for they are found in non-capitalist societies, and in civil society within capitalism. But what of the converse possibility?: can capitalism exist independently of patriarchy, racism etc.? The difficulty here lies in the fact that cases of capitalist institutions which are non-sexist, non-racist, non-homophobic, etc. are extremely rare. Moreover, there are often complex combinations of these structuring forces, in which they adjust to one another (e.g. Cockburn, 1983; Whatmore, 1990). Thus the way in which capitalist firms define jobs and select workers for them is almost universally gendered. Concrete forms of capitalism and patriarchy have adapted to one another, so that the combination in Japanese society is different from that in Sweden, for example.

However, as Walby shows, capitalism and patriarchy conflict as well as combine; for example, sexism can actually make capitalist production more expensive (Walby, 1986; see also Hartmann, 1979). On the one hand, patriarchy makes women's labour power cheaper than it would otherwise be, on the other hand it reduces the size and quality of the workforce available to capital. Similar arguments could be put forward regarding the relationship between capitalism and racism or homophobia. The particular concrete forms that capitalism takes therefore exist sometimes because of these divisions and sometimes in spite of them. In some cases, certain capitalist enterprises would actually cease to exist if the social divisions which they were exploiting were abolished; thus, a clothing sweatshop in London might not be able to survive without access to cheap ethnic-minority female labour, but in that event other clothing firms could survive by other means.[8] Although capitalism

---

[8] An equivalent case discussed by Whatmore (1990) notes how certain kinds of capitalist farming depend on the gender division of labour within farm households. Again this does not mean that capitalist farming in general requires such a gender division in order to exist.

invariably adjusts to these other non-class social divisions, they are not necessary for the existence of capital and class, capital accumulation, etc., as such. Of course, one should expect competition to force capitalist firms to exploit to their advantage every source of inequality they can find, if doing so makes them more profitable, but that does not make those external sources of inequality indispensable to the functioning of capitalist firms in general. If there were no external inequalities to take advantage of, firms would have to resort more to other methods of competition, such as mechanization. Capitalism is highly adaptable. Wage labourers are indeed often men serviced by dependent women in patriarchal domestic relationships, but from the point of view of capital in general they don't have to be; capital can employ men and women with diverse domestic situations. While capitalists and workers may be sexist, racist or homophobic, they may also be feminist, non-racist, non-homophobic or gay, and job specifications and employment policies could be gender-neutral and non-racist. To my knowledge, no one has managed to demonstrate why capital and class, capital accumulation, etc., could not exist in the absence of patriarchy or racism. A non-sexist, non-racist (non-homophobic, etc.) capitalist society would certainly be different from our present society, but there is no reason why capitalist development could not continue (Delphy and Leonard, 1992). It is no answer to say that we would not want such a society (a strange response as it would surely be preferable to our current society). What one would prefer to exist is a separate question from what can and does actually exist. The possibility of a non-patriarchal, non-racist capitalism is not an idle, speculative matter, for it tells us much about the nature of our current society. If we imagine that capitalism can only exist with the support of patriarchy when in fact it can exist without it, then we have seriously misunderstood it. What we have here then, is a false inference from the pervasive and intricate coexistence of class, gender and ethnic differences to the conclusion that none of these sources of difference and inequality can be removed without abolishing the others. The relationship between capitalism as a system and patriarchy or ethnic divisions is therefore contingent. This conclusion demonstrates the importance of asking counterfactual questions. A dual system theory of this kind can acknowledge that in concrete cases, patriarchy and capitalism (or socialism) interact and adjust to one another in complex ways. But since they are separable structures, then *for the purposes of abstract theory* it is possible to theorize gender independently of class, and vice versa. However, the situation for class theories which are supposed to double up as stratification theories is quite different, since they deal with concrete outcomes which are co-determined by other sources of social division besides those of class in the restricted, abstract sense.

Capitalism is easier to condemn if it can be held responsible for as many social ills as possible, and since it invariably exploits any social divisions that exist it is easy to imagine that it cannot survive without them. As we shall see later, there is a common tendency in radical thought to assume that because an alternative society would be best if all the good things went together, they can only go together as a matter of fact. This complements the converse assumption that the social bads can only exist together, not separately or in combination with good things, as in the case of a non-patriarchal capitalism. As we shall see, good and bad very easily go together.

The Significance of an Advanced Social Division of Labour

We now come to the heart of our critique of radical political economy, the treatment of the social division of labour in an advanced economy.

One of the most persuasive features of radical political economy has been its argument that the recurrent problems and crises of capitalism are not mere residual aberrations which can be removed by an appropriate mix of reforms but are endemic to the very structure of capitalism. Thus the inequalities and irrationalities of uneven development in the First World and Third have been linked to their overwhelmingly capitalist nature. In characterizing these structures, it is the diagnostic features of capitalism which have repeatedly been singled out – the capital–labour relation, profit as a regulator of economic activity, capital accumulation. But these are not the whole of the problem. Despite all the attention given to the international and other spatial divisions of labour, little has been written about the economic implications of division of labour. Capitalism has an advanced division of labour in which specialization, fragmentation, interdependence and internationalization have been taken to unprecedented extremes. Though probably not to the same degree, the former state-socialist countries also had a considerable division of labour; the former Soviet economy has been estimated to have produced between 10 and 15 million different types of product, in thousands of different enterprises (Nove, 1983; Rutland, 1985): presumably the totals for advanced capitalist countries are no lower. It is vital to take this complexity seriously if one is to assess the adequacy of actual or preferred modes of coordination.

It is of course capitalism itself which has been responsible for accelerating, deepening and extending the division of labour far beyond anything previously experienced. In so doing, it has created a structure of social relations and activities that has causal powers and liabilities of its own, and which can outlast the overthrow of capitalist social relations

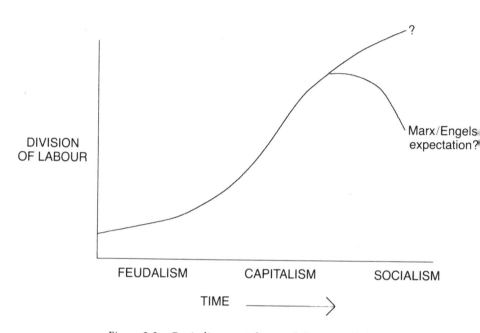

Figure 3.2   Capitalism, socialism and division of labour

of production (figure 3.2). Although, under capitalism, the effects of division of labour always operate with and through other structures and mechanisms of capitalism, they are not reducible to the latter, to the profit motive or class relation, though Marxism typically makes this reduction. The fact that capitalism has deepened this division of labour so much does not mean that the latter has no effects of its own.

The growth of this division of labour is very much part of the development of the forces of production: the latter cannot be divorced from the former. For the reasons identified largely by Adam Smith, division of labour has brought enormous increases in productivity and wealth (Smith, 1976). Marx and Engels do not appear to have dissented from this view, and they celebrated the growth of the forces of production under capitalism:

> How far the productive forces of a nation are developed is shown most manifestly by the degree to which the division of labour has been carried through. Each new productive force, insofar as it is not merely a

quantitative extension of productive forces already known, causes a further development of the division of labour.

(1974, p. 43).

Yet extraordinarily, while they realized that division of labour was a necessary condition of economic development, they also believed in the possibility and desirability of the transcendence and abolition of the division of labour under communism (the Marx/Engels trajectory in Figure 3.2), without this entailing a return to relatively primitive levels of living (See especially *The German Ideology* part 1 in relation to *Capital*: Rattansi, 1982; Selucky, 1979). Like later social theorists such as Durkheim and Tönnies, Marx and Engels recognized that the development of an advanced division of labour necessarily corrodes community or *gemeinschaft* relationships and creates alienation. The expropriation of capitalists would not alter this unless the division of labour were simultaneously drastically reduced. Marx's analysis therefore both confirms and attempts to deny that one could have an advanced economy without alienation. Once again we see that the standpoints of Marxism's critique of capitalism are glaringly contradictory.

This division of labour is not only remarkably complex (Blackburn, 1991a), but intrinsically intractable, and resistant to attempts to socialize control of the economy. This intractability derives from several interrelated aspects:

1. there are far too many producers and consumers for it to be possible to 'pre-reconcile' supply and demand *ex ante* for each product. The diverse, specialized needs and conditions present in an economy of such a scale are unknowable and hence insusceptible to effective centralized control. In the absence of such control, the opacity and indirectness of the relationship between producers and consumers obscures the relationship between the effort put in by the former and the satisfaction of the latter in using their products, thereby creating scope for problems of motivation;

2. advanced divisions of labour are difficult to manage without *ex post* coordination through the price mechanism, thus attempts to replace markets by planning have great difficulty in assessing the costs and benefits of alternative sources of supply and the relative intensities of the host of different demands or needs;

3. the technical incompatibility of many different activities inhibits their combination – while complementary and related products like pulp and paper are susceptible to joint production, non-complementary and unrelated products like air travel and newspapers are not;

4. as we noted earlier, divisions of labour both support and are influenced by the formation of different interest groups and provide them with a basis

for organization. In the case of professions and trade unions, sectionalism and strategies of closure may be more prominent than class solidarity;

5. further, as already noted, the division of labour in the formal economy tends to 'embed', and be shaped by, wider divisions in society, particularly gender and race.

While (1), (2) and (3) derive purely from division of labour, (4) and (5) derive from contingent combinations with other sources of fragmentation. It is important to note that while (5) increases the intractability of an advanced division of labour, (4) would be present even in the absence of (5). This reinforces our point that the division of labour is not to be treated as an inert background against which the principal forces of gender, class and race move; it is an active force in its own right (Sayer and Walker, 1992).

Although some of these points – particularly (1) – are associated with liberals such as Hayek, they are also acknowledged by many economists sympathetic to socialism (Hayek, 1976; Nove, 1983). Moreover they have been confirmed in practice by the failure of central planning in the former Soviet bloc to control more than a small subset of its vast range of products (about 60,000, according to Rutland (1985)). At best, what passed for planning in the Soviet system involved considerable bargaining between the centre, major lobbies of industry and the military (Burawoy and Kritov, 1993). The enormity of the problem of planning a whole economy far outstrips the capacity of any foreseeable computers. The most elementary lesson of socialist economics tells us that the idea of millions of 'associated producers' collectively controlling an advanced economy *ex ante*, without recourse to *ex post* regulation through markets and without needing prices to assess allocational efficiency and relative scarcity, is a quaint pipe dream. That such ideas are entertained illustrates the common tendency to underestimate the complexity of advanced economies.

But how do we explain this anarchy? Why is control of the economy so fragmented under capitalism, with a multitude of capitals pursuing their own goals, and why does coordination between them come about primarily *ex post* through markets? The usual answer is that it is anarchic because control is fragmented amongst separate capitals, each pursuing its own self-interest, with the implicit counterfactual that without capitalist control, coordinated *ex ante* control would be possible. But this is an entirely unsatisfactory and question-begging answer.[9] It is not that capitalist firms, as 'hostile brothers', choose to remain fragmented. Most capitals are constantly trying to expand to control a

---

[9] For example, see Devine, 1988, p. 121.

greater mass of production, which usually means a wider swathe of the division of labour. Why then doesn't each capital control more? Why hasn't the process of industrial concentration continued until it has produced a mere handful of centralized bureaucracies? The answer is not that capitals do not try to reduce the fragmentation of the economy and increase their *ex ante* control of economic processes. Rather an advanced division of labour is too big and intractable, the characteristics and purposes of all the constituent members and activities too numerous and diverse, and the information needs too great, to enable the anarchy to be eliminated without major efficiency losses. Even though producers can plan their supply, they can't plan final consumer demand, and this is not only a macro-economic problem of regulation but a micro-economic one too. Given this relative autonomy of demand from supply, a multiplicity of enterprises can cope with its flexibility better than can a single, centralized supplier. Certainly some kinds of need can be met successfully through central planning, but the idea of a single, collective 'subject of history', successfully planning an advanced economy involving millions of people and their myriad needs *ex ante* (and doing so to the satisfaction of all!) is absurd. Fragmentation and partial submission of the members of the division of labour to *ex post* control is inevitable, and it is so more as a consequence of the complexity and intractability of the division of labour than of the character of the social relations of production within each of its units. (This is not an endorsement of the *status quo*, for their are many as yet untried mixes of ownership, market regulation and non-market control (Nove, 1983; Brus and Laski, 1989; Sayer and Walker, 1992).)

Competition contributes much to the anarchy of the social division and is of course crucial to the dynamics and uneven development of capitalism. But again, contrary to the impression evident in radical political economy, the competitive character of capitalism is not purely a function of capitalist ownership of the means of production. Where there is no possibility of central, coordinated *ex ante* control of a division of labour, there will be decentralization and fragmentation, and where there is decentralized control exercised by producers buying and supplying commodities from and to a common market, competition is almost inevitable, be the producers capitalist firms or worker-managed cooperatives.

In socialist thought, cooperation is often considered to be the opposite of competition, but the social division of labour, stretching globally across producers and between producers and consumers, is too large, too complex and too diverse to allow cooperation. And in any case, as John Stuart Mill pointed out in the mid-nineteenth century, the opposite of competition is (also) monopoly (Mill, 1970). Marxism's emphasis on people as producers tends to give it a somewhat one-sided view of cooperation, as that which might exist between workers who at present have to compete under capitalism, rather than cooperation between producers and users. Producer cooperation tends to lead to the creation of a monopoly of producers over users. Interproducer conflict is reduced at the expense of creating increased scope for producer–user conflict – a reversal of the situation under capitalism – not a suspension of economic competition and antagonism but a restructuring of its principal axes. To reply to this by saying that workers are also consumers and that therefore there need be no conflict of interest simply misses the point of a division of labour; precisely because each worker produces only a handful of different kinds of commodities or just contributes to the production of a single commodity, and yet consumes hundreds or thousands of different commodities, there is ample scope for producer–consumer/user interests to diverge.

That anarchy and competition are not simply a consequence of capitalist ownership of the means of production can be seen by considering what happens when workers take control of the means of production and abolish class. If they also try to abolish the anarchy of the division of labour through central planning, then: (i) they find themselves under the dominance of a quasi-class of planners and bureaucrats responsible for the plan; (ii) they suffer the inefficiencies of the massive inherent coordination and motivation problems so fully examined by students of state socialism (e.g. Aganbegyan, 1988; Brus and Laski, 1989; Kornai, 1986; Lavoie, 1985; Nove, 1983; Putterman, 1990; Rutland, 1985); and (iii) they lack even the limited *ex post* influences upon the economy and its directors afforded by market coordination (producers have to respond to the plan, not to consumer demand). In effect, as is evident in early socialists' talk of society as 'one big factory', an attempt is made to extend the despotic *ex ante* control of the factory to the economy as a whole, thus abolishing the social division of labour as defined above. As we shall see, the failure to carry this through in practice tells us something about the limitations of Marx's definition of the social division of labour and his explanation of its 'anarchy'. If, on the other hand, the workers establish a kind of market socialism in which they own the enterprises in which they work (but not those of other workers), and live off revenue from producing commodities, then they have abolished class without taking control of the social division of labour, so that they are still subject to its 'anarchic' market coordination. Of course there are many other variants of socialism but the above ideal types establish the limits to the distribution of economic power set by the twin axes of class and an advanced division of labour.

While I am arguing that an advanced division of labour is intractable whatever the social relations of production, I do not of course wish to deny that the latter make a significant difference. Capitalism and market socialism as defined above both use *ex post* market coordination but there are significant differences as well as similarities in their behaviour. Worker-owners need not attach the same significance to profit; investment funds would be less mobile as worker-owners would have no interest in giving up their jobs so that lower paid workers and more lucrative opportunities could be employed elsewhere. However, under both systems, enterprises and workers are vulnerable to bankruptcy and unemployment, and macro-economic problems are similar. So although I wish to upgrade the significance attached to division of labour and problems of its coordination, I have no intention of abandoning concerns with class and ownership.

Figure 3.3 sums up these dilemmas. It illustrates the relationship between power in terms of class and in terms of control over the division of labour by reference to ideal types of four different forms of organization of advanced economies – which necessarily means economies with advanced divisions of labour. On the vertical axes we distinguish class societies, in which production and the distribution of the surplus is determined by a minority, from classless societies in which ordinary producers themselves have significant control over these. The horizontal axis ranges from situations in which control over the different parts of the division of labour is fragmented and hence regulated *ex post* largely through markets, to situations where it is highly centralized and regulated *ex ante*. There are two infeasible zones: that on the left indicates the impossibility of an advanced economy without some degree of centralization of control, for example, for providing infrastructure which private enterprises or cooperatives need but cannot provide themselves; the zone on the right indicates the impossibility of total *a priori* control of a complex, advanced economy. This zone broadens in the absence of a ruling class because of the impossibility of planning a complex modern economy without considerable centralization of power.

Capitalism is both a class system and one which is coordinated largely *ex post* through markets. Socialist revolutions generally try to abolish both these characteristics by establishing a classless society coordinated not by markets but by democratic planning. They therefore head southeast on the diagram where they find that the *ex ante* control of an advanced economy is only partly feasible, and then only by centralizing control so that the various parts do not pull in different directions. The need for central control requires planners and a ruling bureaucracy which become a new kind of ruling class or ruling quasi-class, because

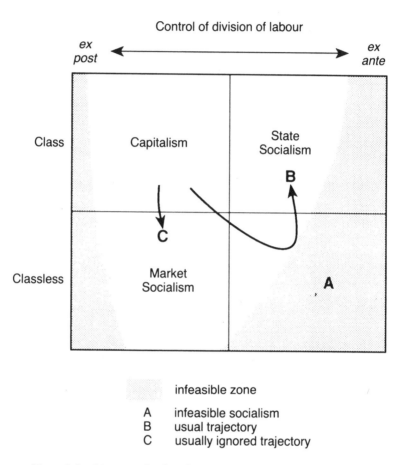

Figure 3.3   Limits to the distribution of economic power under an advanced division of labour

of their ability to control the surplus. The revolutionary trajectory therefore turns northwards, abandoning what I have termed 'infeasible socialism'. The other ideal-typical option, termed market socialism, consists of commodity producing, worker-owned cooperatives competing with one another in markets. It therefore abolishes class but retains the anarchy of the social division of labour which was so central a target of Marx's critique of capitalism (cf. Przeworski, 1991).

I will comment on these different kinds of economy later, but for now, the point of these ideal types is to illustrate the interdependencies between class and division of labour in advanced economies in particularly stark fashion. The only conceivable way out of the dilemmas involved in these relationships is through a drastic reduction in the division of labour and hence a shift to a lower level of material development. Actual histories of socialist economies are of course more complex than our stylized account suggests, and in both capitalist and socialist economies there is always scope for mixed solutions, indeed they are necessary (Hodgson, 1988). Some activities may need or lend themselves more easily to centralization or to democratic control than others (e.g. railways compared to plumbing) (Nove, 1983). Planning may be merely indicative instead of imperative. Notwithstanding these complexities, figure 3.3 still depicts the basic structure of the distribution of power in relation to class and division of labour.

Economic Power, Scale and Scope

These points can be reinforced via a different route, using the concepts of economies and diseconomies of scale and scope from industrial organization theory. There they have been widely used for explaining industrial structure, in particular the diverse combinations of markets and hierarchies (firms) in different industries (e.g. Scott, 1988; Storper, 1988). Since these concepts help explain how many activities each firm can control they can also tell us much about the intractability of complex divisions of labour and about the distribution of economic power. The power of a firm derives not simply from its ownership of means of production but from its ability to combine activities, within and beyond its boundaries. Economies and diseconomies of scale and scope involved in combining a variety of activities are therefore crucial. They derive from two sources:

A: the powers and liabilities or limitations of specific forms of organization, be it a multi-divisional capitalist firm, a flexibly specialized network of small firms, a government department or whatever.
B: the powers and liabilities of the activities which the organization tries to control, be they those of producing bread, missiles or running a hospital or crèche.

A and B can be changed, but in accordance with the causal powers and liabilities of their constituents; neither is completely plastic or capable of doing just anything. In Marxist terms, A and B are equivalent to the social relations and forces of production respectively. Note that the latter term needs to refer not just to a certain level of development of technology, know-how and skill, as if it were merely a historical marker, but to these things in their spatial, horizontal aspect, as they are increasingly differentiated and divided into specific economic activities, and distributed across geographical space.

Naturally, A and B interact recursively, the activities being shaped somewhat by the modes of organization, and vice versa. But this does not mean we can reduce one to the other. The most common fault here is a kind of institutional voluntarism in which B is reduced to A so that activities then appear infinitely plastic in the hands of the organization. Certainly, steel, clothing, insurance policies or foreign holidays (B) can each be produced within a variety of forms of organization (A). However, the economic effects of doing so will differ, precisely because they are materially different activities, which is largely why their industrial organization tends to differ.[10]

Diseconomies of scale and scope limit the power of capitalist firms to expand their internal technical divisions of labour, over which, as Marx put it, they have 'despotic' control, at the expense of the anarchy of the social division. Talk of a 'monopoly capitalism' in which leading firms have already made inroads into the social division is singularly inappropriate, for although the firms may continue to grow, so too do the markets in which they operate, partly as a result of the actions of the firms themselves. Markets and enterprises do not have to grow at each other's expense (Auerbach, 1989; Auerbach et al., 1988; Lindblom, 1977). If there are now fewer and larger firms than hitherto (a disputable assumption), competition shows no sign of diminishing, for the so-called era of competitive capitalism was one of local or national spatial monopolies, in which each firm had few competitors; by contrast the multinational giants of today are exposed to more sophisticated competitive strategies on a much larger scale.[11]

The upshot of these points is that simply to blame private ownership of the means of production and profit maximization (a particular form of organization) for the anarchy of the social division of labour and hence to imagine that eliminating capitalist ownership would also eliminate anarchy and the associated uneven development and economic irrationality, is to fall victim to the voluntarist or idealist fallacy of reducing B to A. Similarly, to attribute the anarchy to the lack of planning – as if only the form of organization (A) and not what had to

---

[10] Note, incidentally, that the reduction of B to A is equivalent in the realm of philosophy to an idealist epistemology. The contrary error of reducing A to B, so that forms of organization are uniquely determined by the material properties of that which they organize, is equivalent to vulgar materialism, while recognition of both the interdependence and the irreducibility of A and B is consistent with critical realism.

[11] As Auerbach shows, the degree of competition in an industry cannot be inferred from industrial concentration ratios (1988).

be organized (B) mattered – is to suggest, against considerable evidence, that successful comprehensive central planning of an advanced division of labour is possible – in other words, that the intractability of the division of labour (B) at the social level can be overcome (Brus and Laski, 1989). Likewise the Marxist notion of coordination of an advanced economy by the 'associated producers' is idealist in both senses of the term. We need a more materialist view of the division of labour and forces of production. The difference between technical or detail divisions of labour within enterprises and the social division of labour between them is not purely one of ownership and despotic versus anarchic control; it is also influenced by the material properties of the different activities requiring coordination. Some combinations yield easily to *ex ante* control, others only at high cost. The extent to which any individual, group or class can control its economic circumstances is therefore not only a function of its ownership or non-ownership of the means of production, but of the extent to which it can control the division of labour, whether across a small part or across vast swathes of it.

But on closer inspection, the relationship between class and division of labour is more dialectical than this, though they still can't be reduced one to the other in either direction. This can be shown in relation to the nature of ownership and economic power. Legal definitions of owner-ship of means of production settle on paper that which is usually contested to some degree in practice. Thus, as industrial sociology shows, the powers that accompany ownership of a firm are limited by the powers of those within it (who may pursue self-interest, gender, class, occupational, or race-related interests, instead of company interest). And as industrial organization theory shows, a firm's powers depend also on the situation outside its boundaries – customers who socially validate its power through their spending patterns, suppliers, creditors, etc. In these ways, the 'vertical' power relations within the enterprise are strongly mediated by 'horizontal' relations across the division of labour, both within the technical division of the enterprise and outside in the social division. In other words, in practice, economies and diseconomies of scope help to delimit ownership and class power, and vice versa.

The Technical and Social Divisions of Labour Revisited

The idea that economies and diseconomies of scope shape the develop-ment of the technical and social divisions of labour takes us beyond the terms of Marx's conceptualization of that distinction, exposing its lack of consideration of the material side of the organization of modern economies. At this point I will therefore try to clarify the shortcomings of Marx's treatment of this subject.

It is vital to note that the distinction as drawn by Marx relies on not one but several criteria. This is to be expected, as he uses it to analyse capital at a fairly concrete (many-sided) level. However, he fails to note that the criteria are separable and need not correspond. In other words, this is an example of his underestimation of the presence of external relations and hence the possibilities of a disaggregative analysis. Four criteria can be identified for distinguishing the technical and social divisions of labour respectively:

1   divisions of labour which are regulated by *ex ante* control, versus divisions which are regulated *ex post* through markets in terms of changes in prices and in quantities produced and sold.
2   divisions of labour under single control or ownership, versus divisions under the control of different, independent owners. While Marx assumes that the owners concerned must be capitalists, they don't have to be; the criterion could apply equally well to a market socialism of worker-owned commodity-producing cooperatives.
3   the division between workers performing different tasks in the production of a single commodity versus that between workers producing different commodities.
4   more implicit is a distinction between divisions of labour whose shape is primarily a reflection of technical characteristics of production, versus divisions whose shape is primarily determined by the relative quantities of different commodities consumers want to buy; for example, the relative numbers of people required to make a commodity and needed to do the paperwork on it depend largely on the technologies of manufacturing and handling invoices, etc., but there is no such technical relationship between the number of newspapers and holidays sold, and hence between the numbers of newspaper workers and tourist-industry workers.

While Marx developed his distinction in order to analyse capital and capitalism, the criteria can also be applied to other forms of social organization of production in advanced economies. All four criteria could apply to market socialism, as defined earlier, if under (2), we define the members of a cooperative as being co-owners rather than separate owners. Under centrally planned state socialism, there is an attempt to abolish the social division of labour in terms of division between separate owners (2) and to replace market regulation with planning (1). This represents an attempt to extend the technical division of labour to the level of society. Remarkably, despite characterizing the technical or detail division as despotic, Marx did not appear to see this as a problem.[12] However, state socialism is or was still stuck with a social division of labour in terms of (4), and to a certain extent (3), a fact of immense significance for its performance.

As research on industrial organization has made clear, (1), (2), (3) and (4) need not always correspond under capitalism either. Where a single firm makes several different commodities, the division of labour between them depends on market conditions. In these circumstances, a multi-product firm could be said to internalize some of the social division of labour, according to criteria (1), (3) and perhaps (4), but not (2). Conversely, where separately owned firms within a production chain collaborate and engage in relational contracting, a degree of *ex ante* control can emerge so that some of the characteristics of the technical division of labour can exist between firms, according to (1) and (4), but not (2), and not (3) (since they still sell each other commodities). That the four criteria for the technical/social distinction do not always match up should hardly surprise us, for capitalist firms are obliged to search continually for new ways of coping with divisions of labour. Industrial organization theory shows us that it is possible for the boundaries of firms to be drawn in many different ways with respect to the material organization of production and exchange, but not just in any way; if the material character of what is being organized had no effect, industrial organization would be random.

Of the four criteria, (4) might appear to be of least interest to social scientists, for it largely concerns a material distinction, between activities involved in producing technically related products and services and activities involved in producing technically unrelated activities. In some cases the distinctions of control, ownership and commodity form, identified by (1), (2) and (3) respectively, do not appear to have much to do with material relationships among the range of activities. For example, there seems little material difference between making the various parts of a computer in several separate firms, where they become separate commodities before being assembled into the final product, and making and assembling them within the same firm. It seems that the

[12] Marx writes: 'The same bourgeois consciousness which celebrates the division of labour in the workshop, the lifelong annexation of the worker to a partial operation, and his complete subjection to capital, as an organization of labour that increases its productive power, denounces with equal vigour every conscious attempt to control and regulate the process of production socially, as an inroad upon the freedom and the self-determining "genius" of the individual capitalist. It is very characteristic that the enthusiastic apologists of the factory system have nothing more damning to urge against a general organization of labour than that it would turn the whole of society into a factory' (*Capital*, vol 1, p 477). Here Marx scores an embarrassing 'own goal', for it is all the more hypocritical for a critic of the factory system to recommend scaling its despotically controlled technical divisions of labour up to the level of the whole society.

tasks of making parts and of assembling them might equally be done by a single firm or by separate enterprises, so distinction (3) might therefore seem arbitrary and asocial. The computer example is persuasive as an example of the weakness of criterion (3) because it concerns closely related activities, among which there are economies of scope. But consider the relationship between computer production and cheese production. The relationship between making cheese and assembling computer components is obviously different from that between making cheese and packaging it or that between making the computer sub-assemblies and assembling them to make computers. In the cheese and computer example the material distinction represented by (4) cannot be dismissed. The rationale of combining computer sub-assembly production and computer production is of a different order from that of combining either with a stage of cheese production. Even if the relevant means of production were owned by the same body, so that they formed part of the same technical division of labour under criterion (2), there is no reason for coordinating production of cheese and computers, for thanks to their lack of relationship they cannot belong to the same technical division of labour under criteria (1), (3) or (4). It is not only that they have nothing in common as regards their production, but that the consumption and hence demand for them is unrelated. Whereas the demand for computer sub-assemblies is derived from that of computers, there is no such relationship between the demand for cheese and computers.

The apparent subtext or subconscious hope of recent radical literature on industrial organization, particularly regarding 'flexible specialization' and 'post-Fordism', would seem to be that while we might have to abandon the hope of centrally planning a whole economy, an allegedly new kind of relationship intermediate between planning and market regulation called networking or relational contracting is beginning to take over coordination of the economy. There may indeed be an increasingly technical character to interfirm relationships within production chains producing related products – that is, technical in the sense of criterion (1), as some element of planning or at least non-market negotiation enters these relationships – but this is not generalizable to the relationships between products belonging to different production chains.

In practice, therefore, we find further confirmation that while industrial organization can take many forms it can never be indifferent to the material properties of the activities that have to be controlled. The characteristics of the social division of labour are therefore not simply the product of a contingent form of social organization or system of ownership. It has an irreducible material character that cannot be entirely suppressed by different forms of ownership and organization. As Selucky puts it, 'No legal, political or organizational change is able to abolish the social division of labor or to turn the division of labor into that within the enterprise' (1979, p. 33).

In some respects, particularly with regard to the technical division of labour, Marx is aware of the material constraints on social organization. Thus, in discussing authority and supervision within capitalist workplaces, Marx argues that they have a 'double nature', deriving in part from the technical imperatives of close coordination, discipline and hierarchical control in large-scale production, and in part from the capitalist ownership of the means of production. Marx envisages the possibility of this latter 'capitalistic shell' splitting off from the technically determined relations of the division of labour within large-scale production units (Marx, 1972, p. 387). Hence, non-capitalist, large-scale production would still need cooperation, discipline and some degree and form of hierarchical control, though not necessarily of the same type as found in capitalist enterprises.

This is an invaluable distinction for thinking about authority and discipline in capitalism, but it is a great pity that he does not apply the same kind of distinction at the level of the social division of labour. At one point in his discussion in *Capital* he refers to it as the 'totality of heterogeneous forms of useful labor, which differ in order, species and variety' (vol. 1, p. 132) – a decidedly materialist definition. However, for the most part, he concentrates on its social characteristics (criteria 1 and 2 above). He therefore failed to see that the organization of an advanced economy and its associated distribution of power and modes of coordination is not only a function of the capitalist form of ownership of the elements of the social division of labour, but also of the material character of an advanced division of labour itself. Just as a large-scale production unit necessitates certain kinds of social relation of discipline and coordination, so the vast range of different production units in a social division of labour requires different forms of organization. Consequently, in his fragmentary comments on socialism and communism, Marx fails to see the material divisions of the social division of labour as an obstacle to the socialization of the means of production. If he had done so, however, it would have had fatal implications for his romantic vision of communist society. The obstacles to the 'socialization' of control and power in the economy are therefore not only the current capitalist social relations of production but the nature of what has to be controlled. For Marx, the development of the forces of production under capitalism brought closer the possibility of socializing control of the economy under the collective will. But if we remember that the development of the forces of production involves not just a rising level of technology, science and skill and increased potential to overcome scarcity, but a development horizontally and spatially of an ever more differentiated and extended division of labour, then we can see that the possibilities for socializing control actually recede rather than increase.

To accept these double-nature arguments is not to rule out all possibilities of increasing accountability and popular control. To make that inference would be to suppose that capitalist enterprises and the wider economy only had a single, technical nature. Capitalist social relations are still recognized as influencing the nature of enterprises and the wider economy in ways which restrict the distribution of power and distort the economy, though their effects are not as sweeping as many Marxists assume. Consequently it does not rule out the possibility of a post-capitalist, non-state-socialist, advanced society which could counteract these effects, even though it could not do much to reduce hierarchy in organizations and reliance on *ex post* regulation of the wider economy.

However, accepting the double nature arguments does not entail accepting even this combination of hierarchy and anarchy. It is still open to us to say: we recognize that large-scale production of good *x* is impossible without a highly hierarchical form of organization (whatever the social relations of production), but that being the case, we would rather do without so much of good *x* in order to avoid such hierarchy. More generally, we could accept that an advanced economy entails fragmented *ex post* control, but argue that it would be better to forgo the material rewards of such an economy and move to a simpler, more primitive economy, for the sake of being able to have more *ex ante* control. In other words, while my argument so far lends some support to a kind of 'industrial-society thesis' in which certain broad features, that radicals conventionally treat as specific effects of the contingent capitalist social organization of advanced economies, are in fact partly inevitable in any advanced economy, it does not follow that we must accept advanced economies, warts and all. One could equally use the argument to make a case for returning to a lower level of development.

Hayek's Catallaxy and Economy

A further source of support for our thesis of the intractability of the social division of labour comes from Hayek's important distinction between 'catallaxy' and 'economy' (Hayek, 1988).[13] Hayek uses 'econ-

---

[13] As the 'father' of the New Right, the name of Hayek may alarm some readers, but I think some of his concepts are useful and do not actually entail the right-wing conclusions he attempts to draw from them.

omy' (hereafter 'economy*') in a restricted sense, close to that of its original Greek meaning, to refer to clusters of economic activities which are organized for a specific purpose and have a unitary hierarchy of ends, and in which knowledge of how to achieve these ends is shared; for example (arguably!), households and firms. By contrast, a catallaxy has no unitary hierarchy of ends, but a mass of innumerable economies* without a specific, common purpose. It is the product of spontaneous growth rather than design.[14]

Not only has the catallaxy evolved without any design, it also eludes any attempts to replace its own market regulation by central control. Interestingly, Hayek interprets this intractability in terms of information and knowledge; an advanced division of labour is associated with an unprecedented division of knowledge, which far outreaches the comprehension of any single mind or group.[15] Knowledge in a catallaxy is not merely contingently dispersed because of fragmentation produced by the self-interested behaviour of many capitals: rather it necessarily eludes central appropriation because of the extraordinary division of knowledge in an advanced economy.[16] While private property is not the cause of catallaxy, it is, for Hayek, 'generally beneficial in that it puts control of individuals' circumstances in the hands of those who know best about them, the individual, rather than in the hands of some other controllers, for however benign their intentions, they will and can know little of those circumstances' (Hayek, 1988, pp. 77–8).

In economies*, choice is key, in catallaxy exchange. This is an important observation for it challenges the assumption of many liberal economists that economics concerns choice and maximization. This may be so at the level of economies*, but at the level of catallaxies, the vast diversity of individuals and institutions across an advanced economy each with different abilities, desires and information, means there is no common standard of merit for judging between the different opportunities open to individuals (Hayek, 1988). My needs and objectives are

[14] Hayek assumes that a catallaxy is coordinated by markets. However, since attempts may be made to coordinate a catallaxy through central planning, and since, as Hayek notes, this cannot succeed in changing them into an economy*, we would argue that such economies have a 'repressed catallaxy'. Moreover, a society of self-sufficient, independent producers could have a multitude of independent purposes and hence be a catallaxy. Hence catallaxy can be considered separately from markets.

[15] Compare Durkheim: 'as a consequence of a more advanced division of labour, each mind finds itself directed towards a different point on the horizon, reflects a different aspect of the world and as a result the content of men's minds differ from one subject to another' (cited in Abercrombie *et al*, 1986).

[16] This point is borne out from the standpoint of economies* (or the technical division of labour) by the Penrosian view of firms, which sees them not merely as artefacts of private ownership but as having specialized stocks of know-how (Best, 1990).

not those of a pensioner, a farmer, ICI, the local police force, a worker in Malaysia or Spain, though I am involved in the same world economy as all of these. This is why attempts to plan catallaxies centrally, using programming methods to optimize the allocation of resources, run into the problem that there is no obvious objective function to maximize (Nove, 1983). Thus for Hayek, the fundamental economic problem is not calculational but epistemological (Kukathas, 1989) – how to coordinate the actions of innumerable agents without the possibility of any adequate centralized knowledge of their needs and resources. In this light, competition operates as a discovery procedure, rather than a solution to cost minimization of a known problem, and the main role of markets is in generating information through the price mechanism as to how economic agents ignorant of each other may best attain their equally unknown purposes. This is not to deny that markets tend to encourage efficient uses of resources; although there may be no universal standard against which to evaluate economizing in a catallaxy, as long as resources are scarce and actors have limited budgets and alternative ways to spend them, then prices will have a rationing effect.

The economy*/catallaxy distinction bears an obvious, if rough, resemblance to the Marxian distinction between the technical and social divisions of labour. To his credit Hayek does not reduce the distinction to a matter of ownership. As we have seen, though it is common for the technical/social distinction to correspond respectively to that between control by single owners and control among many separate owners, the correspondence is not unique or necessary. Hayek recognizes the material and informational differences between the two, independent of ownership, which prevent the successful extension of the principles governing economy* (technical division of labour) to the level of the catallaxy (social division of labour).

While the Hayekian concept of the division of knowledge is a useful one, it only addresses knowledge which is limited to particular specialist activities, as if all economic knowledge were split into fragments to which the scope of enterprises neatly corresponded. To the extent that divisions of labour do give rise to and depend on specialist knowledge, the concept of division of knowledge is useful. But given Hayek's individualism and his rejection of the category of the social, one is inevitably suspicious that more is involved than this, and that he is introducing a view of knowledge which negates its intersubjective, relational dimension. (Conversely, it would be equally fallacious for socialists to suppose that just because knowledge is inherently social, it is only contingently divided, with the implication that a socialist order could replace catallaxy with a single economy*.) Clearly, there is a sense in which knowledge is social, but there is also still a remarkable division of much economically relevant knowledge into arcane specialisms. But precisely because knowledge is divided, information exchange is needed and hence there must always be considerable communication across the division of labour between the fragments. Hayek and other market romantics believe that where markets exist, price information is all that is needed for this purpose. This is absurd: as we shall see in the next chapter, vast numbers of sales and marketing people are needed in market economies to exchange the information necessary for organizing production and distribution. It should not be surprising that while there is a division of knowledge, the information needed for production in a division of labour does not fall neatly within the confines of particular enterprises, or reduce to price information.

The depiction of the distribution of knowledge and goals in society implied by the distinction between economy* and catallaxy is therefore too stark and simplistic. In between economy* and catallaxy there are what might be termed 'proximate economies*,' that is activities which have different kinds of know-how and ends but which are complementary, and for which mere market coordination is insufficient, or inferior to combinations of markets with networking and collaboration. For example, with the development of information technology, many firms making computers and telecommunications which formerly functioned purely as separate economies* have become proximate economies*, needing to share information and develop new products together. We could go further and develop still more differentiated pictures of the distribution of knowledge: a few kinds of economic knowledge must be shared by everyone (e.g. regarding money); some types are restricted to certain spheres which cut across sectors and localities (e.g. knowledge of accountancy); some are restricted to sectors or localities; some are restricted to particular institutions or enterprises, and some to individiuals. Nevertheless, as always, to show that a conceptual distinction fails to acknowledge other kinds of difference is not to say it fails to identify any significant contrast. Although the distribution of knowledge and information is complex, the distinction between economy* and catallaxy remains vital for understanding advanced economies.

In his description of the 'extended order' in a catallaxy, Hayek stresses its complexity, internal diversity and spontaneous, evolutionary development. This makes its diverse qualities unknowable to any single mind or organization. While the system arguably enables individuals or particular organizations to achieve their ends subject to their compatibility with those of others, for Hayek – echoing Adam Smith – it is supposedly futile, naïve or dangerous for them to try to plan or successfully intervene in some aspect of the development of the wider system in order to change it. Such is its complexity and diversity that any actions attempting to achieve general results are likely to be thwarted by the system's counterintuitive qualities.[17] For Hayek there is both a scientific and a moral point here: not only is the knowledge of the would-be planner likely to be inferior to that of specialized individuals, but intervention would constitute an invasion of the latters' liberty. (We will reserve comments on the moral arguments for later.)

Constructivism

The alleged properties of catallaxies are clearly threatening to socialist hopes for a collectively controlled and planned advanced economy. Hayek regards such aspirations as based on a 'fatal conceit', which he terms 'constructivism': intellectuals' activities in contemplating social wholes and seeking rational solutions to problems tempt them to suppose that what can be comprehended in thought, through abstraction, can be rationally and consensually coordinated in practice. This of course is what the Enlightenment project presupposes. To a limited extent, their efforts may be successful:

> Man's [*sic*] inventiveness contributed so much to the formation of super-individual structures within which individuals found great opportunities that people came to imagine that they could deliberately design the whole as well as some of its parts, and that the mere existence of such extended structures shows that they can be deliberately designed.
>
> *(1988, p. 75)*

When this outlook is combined with a preference for collectivism, the tendency towards constructivism is all the stronger. This is nowhere more clear than in the early socialist aims of an economy run as 'one big factory'. Hence Hayek is completely justified in arguing that socialists see catallaxy as a potential economy* whose realization is blocked by its capitalist character and, in some versions, by the lack of development of the forces of production (Hayek, 1988).

Marx and interpreters such as Mandel try to justify their constructivism by arguing that the development of cartels and ever-larger capitals creates 'closer and closer bonds of objective cooperation between producers who are still living hundreds if not thousands of miles apart' (Mandel, in Marx, 1976, p. 946), and replaces market regulation by

---

[17] Note the similarity to the arguments of another right-wing theorist of complex systems, Jay Forrester, the originator of computer models of industrial, urban and world dynamics (Forrester, 1972).

planning, hence constituting 'the abolition of the capitalist mode of production within the capitalist mode of production itself' (Marx, 1972, p. 438). Thus the rationally planned order within the enterprise becomes a draft meaning for a potential rational order at the social level. Similarly Callinicos argues that democratic control by the associated producers is prefigured in the control exercised by workers in factory Soviets (Callinicos, 1991, pp. 111–13), while Gramsci argues 'the socialist state already exists in the institutions of social life characteristic of the exploited working class' (cited in Callinicos, 1991, p. 113).[18] Such positions completely fail to grasp the fundamental difference between running a technical division of labour for producing a particular type of commodity and coordinating a social division of labour or catallaxy involving millions of different commodities, thousands of different enterprises and billions of consumers.[19] The idea that the most fundamental contradiction of capitalism is that between the socialization of production and the private ownership of the means of production is extraordinary. To believe it one has to conjure away the material nature of the social division of labour and its associated division of knowledge. As Callinicos candidly later notes, 'much more thought needs to be given to the nature of democratic forms of planning involving horizontal connections between different production units and between producers and consumers'! (1991, p. 124).

Socialists tend to 'assume that it is unreasonable to follow or do anything blindly (i.e. without understanding) and that the purposes and effects of a proposed action must not only be fully known in advance but also fully observable and maximally beneficial' (Hayek, 1988, p. 71). But the absence of a planner or designer does not entail an absence of order: just as languages are ordered but not the product of any design, so catallaxies are unplanned but ordered. Hayek regards as 'naive' those who 'can conceive of order only as a product of deliberate arrangement' (ibid. 76). Furthermore, for Hayek, as for Smith, the results of attempts at design or intervention are invariably inferior to the workings of the invisible hand of 'the market' (Smith, 1976).

---

[18] Gramsci continues in highly constructivist vein: 'To link these institutions, coordinating and ordering them into a highly centralized hierarchy of competences and powers, while respecting the necessary autonomy and articulation of each, is to create a genuine workers' democracy in effective and active opposition to the bourgeois State, and preparing to replace here and now in all its effective functions of administering and controlling the national heritage' (1910–20, p. 65).

[19] Having mis-specified the distinction between technical and social divisions of labour, and attributed the fragmentation of control in capitalism to capitalist ownership, Devine not surprisingly goes on to associate the socialization of production with 'an enhanced [*sic*] role for information, conscious coordination, flexible adaptation, cooperation and creativity' (Devine, 1988, p. 122).

Many socialists respond to these standard liberal claims by arguing that the unintended consequences of the actions of self-interested individuals in markets are not nearly as beneficial as liberals suppose, and that since many of these consequences can be foreseen even if individuals don't intend them, there is a case for intervention or planning. But the Marxist response is more radical than this, and far more constructivist. For Marx the only good order must be the product of conscious collective purpose, a Hegelian legacy of humanity rising to consciousness and control over itself (Cohen, 1991, p. 17). Marx is resistant not only to actions having bad unintended consequences, but to unintended consequences *per se*. Here economic liberalism and Marxism are poles apart. Whereas liberals celebrate the workings of the invisible hand and the unintended but allegedly benign consequences of actions, Marx's critique of commodity fetishism attacks the way in which the circumstances and social relations in which commodities are produced are obscured behind the commodity itself, so that economic processes work behind our backs, the product of our actions and yet beyond our control.

With the rise of a global catallaxy, and the disillusionments and setbacks of the late twentieth century, the enormous presumption of Marxism's extreme form of constructivism has never seemed more misplaced. Faith in the early liberal providentialism may have waned too, but particularly with the demise of state socialism, the idea of an extended order which is the product of evolution rather than design, and perhaps difficult to improve upon by intervention, is bound to gain strength. However, there are several caveats to be made about what might be termed 'Hayek's conceit of fatalism' in his critique of constructivism.

If Hayek's concept of constructivism ought indeed to give us pause, his own use of it is wildly excessive, for he deploys it fatalistically to legitimize liberal regimes and to disqualify any attempts at intervention. But while Hayek's use of his concept is transparently apologetic, we don't have to follow him in this respect. The nature of a catallaxy as a spontaneously evolved extended order of unimaginable complexity might make us abandon the most extreme forms of constructivism in which a universal historical subject produces a society without contradictions, but it does not entail that more modest forms of constructivism would do more harm than good. The fact that unintended consequences of actions are central to the functioning of catallaxies does not warrant a refusal to intervene in them. Intervention may sometimes make things worse but Hayek provides no good arguments as to why this should always be so. In any case, it is contradictory for any social scientist to break entirely with the constructivist element of the Enlightenment ideal, for the study of society has little point if it cannot liberate and empower people at least to some degree. The critical implications of the division of knowledge for constructivism are also exaggerated. Certainly constructivists would do well to consider the likelihood of intellectuals and state officials having inferior knowledge of local and specific situations to those of particular individuals, groups and institutions within the division of labour. But while Smith and Hayek's warnings on this are salutary, they do not prove that the self-interested behaviour of individuals would always have beneficial effects at the social level. What is rational for an individual is often irrational for a collectivity: the unintended consequences of actions can be paralysing, and predictably so.

Moreover, the emphasis on the dispersion of knowledge can easily become an excuse for turning a blind eye to problems whose causes we could discover and act upon, such as the ecological disasters of large-scale drift net fishing or the social problems of child labour. It invites us to condone selfishness and denigrate altruism. Liberals can use their celebration of unintended but beneficial consequences of self-interested action as an excuse for ignoring those consequences which are damaging, so that they may plead innocence even where such outcomes are foreseeable.[20] Hayek's fatalistic view of liberal capitalism in the metaphysical guise of 'the extended order' as the best of possible worlds is not only presumptuous and apologetic, but – despite protestations to the contrary (Hayek, 1960, p. 397) – thoroughly conservative.

Absent from Hayek's image of capitalism as an unimaginably complex mass of individuals responding to one another through markets is any notion of major social structures. There are affinities here with Karl Popper's critique of revolution and his advocacy of 'piecemeal social engineering'. Though not based on such extreme aversion to intervention, Popper's critique suffered from the same neglect of the presence of major, enduring social structures which would be unlikely to yield to piecemeal change (Popper, 1976).[21] But we don't need to invert this emphasis: while modern societies and advanced economies are indeed catallaxies, they are *also* systems with grand structures. Finally, the apologetic dimension of Hayek's work is evident in his disregard of the anarchic and uneven character of capitalist development. Although he rejects neoclassical economics' static-equilibrium framework, his celebration of the miracle of the market simply ignores the temporal and

---

[20] Nevertheless, liberal warnings to workers about the inflationary consequences of their wage demands show that, when its suits them, liberals are quick to criticize action leading to unintended but foreseeable and unfortunate consequences.

[21] In *The Constitution of Liberty*, Hayek cites Popper's piecemeal engineering approvingly (1960, p. 70).

spatial upheavals associated with the creative destruction of capitalism. Hayek's exaggeration of 'order' is the complement of Marxism's exaggeration of 'anarchy'. The coordination of a catallaxy is remarkable – miraculous even – and not to be taken lightly, but in association with capital accumulation, the invisible hand is also extraordinarily uneven and destructive in its operation.

In sum: Hayek's concepts of catallaxy, economy* and division of knowledge are vitally important for understanding advanced economies, and they give a fresh angle on the familiar liberal story of the miracle of the market. However, the implications he draws from these concepts are tendentious and transparently apologetic. I therefore suggest that we do not allow the weakness of these implications to discourage us from using the concepts. Likewise his discussion of constructivism turns what could have been a valuable concept for rethinking socialist assumptions into little more than a prop for conservative apologetics, but that should not discourage us from making better use of the concept.

Conclusions

The lack of control that people have over the economic processes that affect them is due not only to class but to division of labour. How *could* they control the latter? How could individuals participate in and reach agreement about decisions concerning all the thousands of different specialized activities involved in producing what they consume? And why should they want to? One is reminded of Oscar Wilde's jibe that 'socialism takes too many evenings'. I don't want to buy goods produced by workers who are super-exploited or which are ecologically destructive, and I therefore want to challenge the division of knowledge where it prevents us knowing such things. But otherwise, I would rather not have to get involved in decisions over what should be produced and how. But in any case, in a catallaxy that would thankfully not be possible. Even where private ownership in the means of production is abolished, the social division of labour and knowledge can remain.

The 'socialization of production' – one of the most mysterious and underanalysed concepts in Marxism – turns out to have more rhetorical power than content. In failing to distinguish the distribution of power in relation to division of labour from that due to class it encourages the illusion that abolishing class leads automatically to the socialization of control across the social division of labour. The only way the latter could be socialized would be by reducing division of labour to pre-industrial levels, which would imply a huge drop in standards of living, threatening not merely luxuries but what people in advanced industrial societies have come to regard as necessities. Divisions of interest between rival producers and between producers and consumers are not merely an incidental and contingent feature of advanced economies but a necessary one.

One possible reaction to the elevated role of an advanced division of labour argued for above is that I have reified it and treated it technocratically. In reply it should be noted that I have acknowledged that divisions of labour obviously consist of social relations, indeed ones which are likely to be competitive and politicized. To stress the materiality of divisions of labour in no way contradicts this, unless one subscribes to an untenable dualism of the material and the social, as if people and their struggles were not also material. Nor does the argument treat an advanced division of labour as immutable. One could argue without contradiction from the above that it should even be abolished, accepting that doing so would drastically reduce material standards of living.

I have attached more significance to division of labour than is usual in radical political economy, and have argued that its effects – for example, in terms of uneven development and anarchic coordination – are irreducible to those of the particular social relations of production present. This is not to say that the particular character of modes of production is irrelevant – far from it, for the effects of division of labour are always strongly mediated by the social relations of production, modes of coordination, types of surplus appropriation, and much else. The argument challenges the balance of radical political economy's explanations rather than rejects them wholesale, but as we shall see in later chapters, it has unsettling implications for radical ideas about alternative economic systems.

4

Markets and Other Modes
of Coordination

In the last chapter we saw that certain features of advanced economies
were present to some degree in centrally planned as well as capitalist
economies, and derived from the existence of an advanced division of
labour. It is perhaps unusual to discuss the implications of division of
labour without copious references to markets, indeed economists generally move straight from division of labour to markets, as if they were
inseparable, or else treat markets as more fundamental. But while
markets imply division of labour, it is possible to have divisions of
labour, even at the social level, without markets, linked (though not
necessarily efficiently) by other modes of coordination. This helps us
distinguish effects of division of labour from those of markets. In this
chapter we shall examine modes of coordination further, but concentrate
on markets as these are the most theoretically and politically contentious
mode.

In recent years 'the market' has become something of an icon, being
fetishized as having powers and authority of its own, to which people
and institutions must conform (Keat and Abercrombie, 1991). Its very
mention triggers strings of associations, negative or positive according
to one's politics. For radicals, 'the market' may connote capitalism
with its evils of unemployment, major inequalities, selfishness and
alienation, and extremes of wealth and poverty in places like south-
central Los Angeles, Liverpool or Rio. Such interpretations bring in
capitalist social relations of production, neo-liberal kinds of regulation
and many other features which presuppose markets but which are not
actually entailed by them. For the New Right, markets – or rather, 'the
market' – is the essence of any economy and connotes both a sphere
of freedom and a source of ultimate authority. In the former state-
socialist countries markets have been treated as a panacea, with little

evidence of critical understanding, though the experience of these countries is already demonstrating the inadequacy of liberal market theory. Liberal and libertarian celebrations of markets have been extensively criticized on the Left for writing out class and inequality (Schweikart, 1992). But while there is no excuse for ignoring class and inequality in accounts of capitalism, markets can exist outside capitalism, and hence they should not be seen as equivalent (see *Review of Radical Political Economy*, 1993; Estrin and Le Grand 1989; Lipietz, 1992; Roemer, 1992).

Theorists with different political views tend to differ not only over the evaluation of markets, but over their very nature, their properties and preconditions. There are also differences between economic and sociological or anthropological treatments of markets. Thus discussions of issues such as the equity implications of markets, or the concept of consumer sovereignty, tend to involve different implicit views of what markets *are*. If we are to avoid confusion we need to discuss the latter question first. The aim of this chapter is therefore to analyse concepts of markets and other modes of coordination and to negotiate a way through the polarized views currently available, so as to prepare the ground for addressing major theoretical and normative issues about markets in the next chapter.

It is not just that concepts of markets differ and that views about markets tend to be polarized, but that discussions are plagued by imprecision and sliding between different concepts, hidden by the use of the same terms. As a result, disputants frequently talk past one another. Some concepts of market are abstract, some more concrete; some are restrictive, dealing only with exchange, others inclusive, extending beyond exchange; some are concerned with latent or imaginary markets rather than actual ones; all vary according to the wider theoretical frameworks or 'optics' within which they are set. While I am critical of many uses of the term 'market', I shall argue that most of the concepts have *some* reasonable 'home sense', but that the common tendency to extend them further is problematic.

Other modes of coordination are common in advanced economies, such as planning, networks and democratic control, but these tend to evoke less controversy, perhaps because they involve intentional control rather than a control mechanism that works via the largely unintended consequences of economic action. Nevertheless, there are some common misunderstandings of these modes which warrant comment, particularly as they are often seen as rivals of markets. In the first section I shall begin by introducing a 'root' definition of markets, and then briefly outline some of their major characteristics and preconditions. The second section presents a conceptual analysis of the many uses of the terms

· 'market' or 'the market'. The chapter ends with a discussion of non-market modes of coordination.

Before we can proceed, there is one kind of conceptual elision which needs comment at the outset. This concerns the difference between 'market economies' and 'capitalism'. Often the former is treated as a euphemism or synonym for capitalism, and frequently that is all it is. Capitalism is a specific, albeit dominant, variant of market economies, one that has attributes – most importantly those to do with the organization of production – which are not reducible to markets. It is important to know whether particular features or behaviours derive from the existence of markets or from those aspects of capitalism beyond the market, or whether they can exist within non-capitalist market economies too. I shall use the terms 'market economy' or 'market system' when I am referring to phenomena that can exist within non-capitalist market situations as well as within capitalism. Further, in order to distinguish capitalist businesses from those of other economic systems (e.g. public companies, cooperatives) I shall reserve the term 'firm' for capitalist businesses. I shall continue to use the term 'enterprise' for units of economic production in non-capitalist economies or where I want to refer to such units in abstraction from the kind of economic system.

Markets in the 'Root' Sense

One of the few theorists to problematize the definition of markets is Hodgson (1988). For him, a market is 'a set of social institutions in which a large number of commodity exchanges of a specific type regularly take place, and to some extent are facilitated by those institutions' (p. 174). A market therefore includes not only commodity exchanges themselves and the associated transfers of money and property rights, but the practices and setting which enable such exchanges to be made in a regular and organized fashion. I shall discuss other usages of the term 'market' later, but for now I want to take this as our 'root' or 'home' sense and enlarge on its implications.

We might add that there is normally an element of competition, with many buyers and sellers involved, and hence the possibility of exit, i.e. withdrawing or doing business with a different buyer or seller. The term 'institutions' is used in a rather elastic way in Hodgson's definition to include not only legal bodies concerned with property rights and contracts, but formal and informal bodies or organized practices for distributing and transporting goods, for setting prices, and for communicating information on prices, products, quantities and potential buyers and sellers to those concerned (Hodgson, 1988). Some might be surprised by the treatment of markets as institutions, since markets are often counterposed to hierarchies, and treated as zones of free association and contract in which, as Hodgson puts it, 'individual preferences collide' (p. 176). However, regular exchanges do not arise spontaneously but require organization and rules, rather than improvisation.

It might nevertheless be objected that some exhanges are indeed made on an *ad hoc* basis, for example, selling a second-hand bike to a friend, but these are unlikely to involve setting prices in competition with other sellers, or to incorporate organized ways of communicating information, and so could hardly be said to constitute a market. In some cases, the organized character of the market is quite clear, especially where it has a specific location, like a stock-market or second-hand car auction. Other markets are organized more loosely, in that although buyers and sellers might only enter the market intermittently, there are organized information channels and procedures by which they can be brought together. Thus, people may only want to buy or sell second-hand cars occasionally, but in addition to auctions, there are specialist trade journals in which they can advertise and which effectively create a competitive market amongst their readers and advertisers.

Markets have become considerbly more organized and routinized as capitalism has developed. Transactions in organized markets are enormously more efficient than atomistic, *ad hoc* exchanges and have been a necessary complement to the development of large-scale production (Hodgson, 1988, p. 181). Thus, Chandler's analysis of the rise of large corporations in the United States in *The Visible Hand* is simultaneously a demonstration of the rise of organized, routinized market exchange (Chandler, 1977). Until exchange was routinized, companies could not get the volume and reliability of throughput to warrant large-scale, highly rationalized production. Naturally, the relationship worked in the converse direction too, with market organization requiring continuous throughput to be worthwhile.

Non-market exchange

Many exchanges of central importance in advanced market economies are not made on an openly competitive basis, usually because there is only a single supplier or buyer, and the costs of finding others or creating new sources of supply would be prohibitive. Thus, a defence firm might buy a particular kind of radar from a single supplier, without considering offers from other firms, on the grounds that its supplier is the only firm which is competent in the manufacture of that item. Trying to persuade others to acquire the specialized know-how and specific assets to make a competitive bid may be a waste of time and money and involve duplication of effort in building trust and exchanging information. Once such purchases have been made, customers can be 'locked in' to making more from the same supplier. Nevertheless, while orders may be repeated for years on this kind of basis, such exchanges are not immune from what are often termed 'market forces' simply because the buyers have to economize and will be reluctant to spend more than necessary, and because there is likely to be at least some threat of actual competition and exit, if the supplier's performance is consistently poor.[1] Moreover, attempting to lock in customers, or trying to unlock those who are already locked in, is a key form of competitive behaviour in itself. Exchanges of this kind are far more common, especially between firms, than much economic theory recognizes. Hodgson (1988, p. 177) terms commodity exchanges made outside markets through some other sphere of activity 'non-market exchanges'. Since, as I have argued, there is usually in practice a possibility of the buyer choosing to exit and 'go to the market', there is a continuum between non-market or non-competitive exchanges and market exchanges. It should also be clear from the defence radar example, that non-market exchange is not necessarily a sign of inefficiency, but may be the most economical way of sourcing specialist products.

Types of market

Markets differ significantly according to the way transactions are organized, particularly with respect to pricing. Spot markets in which prices are flexible, even volatile, and in which relationships between actors are transient, even if repeated, most closely approximate economic models. However, most real markets are not like this. Fixed-price markets in which sellers respond to lack of sales not by dropping prices but by waiting for custom are more common. Fixed prices provide a stable environment for calculating costs and organizing production and distribution. Anything which makes production erratic, whether on the output side or input side, is liable to cause waste and therefore be costly. For this reason, industrial buyers often put reliability and quality of supply before price. There are usually special reasons for the existence of spot markets. Thus, it is not suprising that stock-markets should be spot markets: they offer not produced commodities for sale but claims on future value, and production (whether as productive consumption or production of output) is not directly reliant upon sales of shares. In view of the need for stable conditions for production, we find that many spot

---

[1] In the case of defence contracting, this does not of course apply if procurement operates on a 'cost-plus' basis.

markets are being replaced by more organized markets in which prices are fixed by major actors. Thus spot food markets are becoming marginalized by big supermarkets which organize whole supply chains and have considerable control over prices. The flexibility of prices in spot markets may appear to signal efficient markets but they also reflect poorly organized production and erratic consumption (often as a result of circumstances beyond the control of producers, such as weather).

There is a further significant difference between buying standardized products off-the-shelf and buying goods made to order or by subcontracting, as these represent radically different relationships between production and exchange. Another important difference is between 'arm's-length contracting' and relational contracting. Whereas models of markets tend to assume the universality of arm's-length contracting, in which little information other than price needs to be given, and buyer–supplier relationships are minimal, relational contracting involves information-sharing, trust-building and collaboration between buyer and seller, before, during and after the actual exchange, often even to the point of jointly producing certain goods. To a large extent these different forms of governance or coordination reflect the technical character of the product, but if theories of post-Fordism are correct it would appear that the relational form has recently made some inroads into the domain of arm's-length contracting on account of its superior recognition of the benefits of long-term collaborative relationships (Best, 1990). While the virtues of the latter have been widely noted in the more sociologically influenced literature on industry and markets, the neo-liberal theorists who have the ear of many Western governments seem to imagine that markets work best on the spot, arm's-length contracting model and they actively discourage information sharing beyond what is necessary for the short term.

Information

Flows of information are vital to the operation of markets, but the perfect knowledge assumed in the obfuscatory model of perfect competition is rarely even approached. Moreover, as Richardson (1972) has demonstrated, innovation would never occur under perfect competition because the universal availability of the necessary information would prevent any entrepreneur gaining any reward from innovating. Markets normally operate with limited, imperfect information and foresight and rely heavily on trust. As the Austrian perspective of authors like Hayek makes clear, markets allow micro-economic coordination of a catallaxy, which by definition is characterised by mutual ignorance rather than perfect information. Indeed, ironically, if the latter did exist, it would be a good deal easier to replace market coordination by central planning! As Hayek puts it: 'The whole acts as one market, not because any of the members survey the whole field, but because their limited fields of vision sufficiently overlap so that through many intermediaries the relevant information is conveyed to all' (1952, pp. 85–6).

However, Hayek and other liberal economists invariably overestimate the sufficiency of price as a source of information for buyers and sellers in markets.[2] Even where commodities are not advertised, non-price information normally has to be exchanged if commodity exchange is to take place, its volume and importance depending on the familiarity of the product. This information is normally provided free to the buyer. In other words commodity exchange usually requires exchange or transfer of information without payment, outside market exchange: markets are not self-sufficient but require support from other modes of coordination. Where there are gains from sharing specialized knowledge, for example among firms and their suppliers, it may help to provide this free, though it often requires a non-market organization such as a supplier group or state institution to do this. However, while the extent to which markets are self-regulating and save information, communication and decision-making time may be exaggerated by liberal celebrants of markets, they are still likely to be more economical in their information needs than other modes of coordination.

Even where markets do provide adequate information, enterprises may not be able to act upon it in a way which coordinates supply and demand (O'Neill, 1989). Thus, over-production crises may be prolonged by each supplier waiting for others to cut back. The only way to break such a deadlock may be to suspend competition by setting up a recession cartel to share out capacity cuts. Such situations are common in market economies, regardless of ownership; they can happen not only with capitalist firms, but worker-owned enterprises, and state-owned 'national champions' competing internationally.[3]

Labour and markets

The operation of markets is not costless but requires considerable amounts of labour and organization in distribution, information gathering and communication. Both radical political economy and mainstream economies frequently give the impression that products sell themselves:

---

[2] 'Prices and profit are all that most producers need to know to be able to serve more effectively the needs of men [sic] they do not know'. (Hayek, 1988, p. 104).
[3] Witness the crises of over-production in the European steel industry in the early 1980s (Morgan and Sayer, 1983).

in the former, which has no interest in how commodities find buyers, products are simply 'thrown' on the market; in the latter it often seems that a freely operating price mechanism does all that is required to sell them. Neither does justice to the huge numbers of workers involved in buying, marketing and selling. However, this should not allow us to forget their economical qualities in comparison to other modes of coordination such as planning or democratic control. The labour concerned in running markets is primarily organizing activities at the level of particular supply chains and transactions. The market workers are not involved in regulating the vast input–output matrix of the social division of labour or catallaxy in market economies: this occurs through the unintended consequences of myriad economic actions, which both influence and are influenced by market prices.

Market regulation

Mainstream economics propagates the myth of 'free markets' based on spontaneous exchanges between individuals who have something to sell in exchange for what others have. In this myth, the state is cast as a possible source of 'intervention' or 'distortion'. Yet far from being an unnecessary interference, the state is a normal feature of real markets, as a precondition of their existence. Markets depend on the state for regulation, protection of property rights, and the currency. State regulation of markets need not be against the interests of firms; on the contrary, it may be the result of lobbying by major firms. Especially where these firms are 'national champions' in international competition, it is likely to be in the interest of the government to support them. At a more local scale, even fruit and vegetable markets in town centres are politicized, entry and conditions being regulated by local councils.[4] Markets are not only regulated by the state but by major firms and other formal and informal associations in order to restrict entry through strategies of closure, or to enforce technical standards and structure competition (Cawson et al., 1990; Polanyi, 1944; Marquand, 1988; White, 1993). These are not necessarily 'conspiracies against the consumer' as Adam Smith suggested. A decision to make computer products of different manufacturers conform to certain technical standards in order to make them compatible is helpful to users. Food hygiene regulations obviously benefit consumers too; the freedom to buy food infected with salmonella is hardly worth having! Since a food safety scare can damage the sales of innocent as well as guilty suppliers, it is in the interests of all suppliers to form organizations to regulate themselves

---

[4] I am grateful to Balihar Sanghera for information on the latter case.

in order to increase consumer confidence. The same applies to building firms, insurance companies, lawyers and a host of other suppliers of goods and services in markets. Similarly it is in the interests of enterprises to attempt to police and regulate their own suppliers so as to protect themselves from the effects of faulty supplies on their own sales and reputation. In these circumstances, regulations can themselves become a weapon in power struggles between enterprises (Doel, 1994).

Embedding

Contrary to the mainstream, 'undersocialized' view of markets as operating either in a social vacuum or in a social setting which is of no consequence, markets and exchange are always socially 'embedded' (Granovetter, 1985). The concept of embeddedness is somewhat loose, but it includes the ways in which transactions and other economic actions involve and are set within social relations, and presuppose trust and inter-subjective understandings of rules among actors; thus contracts always presuppose a non-contractual basis or background, as Durkheim argued. Some aspects of embedding involve elements which are vital for markets' functioning, such as those which create trust, but others may be more incidental and simply involve other things which actors do while buying and selling. Embedding therefore takes different forms and degrees.

However, the variation is not arbitrary, for under competitive conditions it is subject to a technical and economic logic. Thus the sale of simple, standardized, familiar goods like newspapers needs less embedding than the sale of complex, costly, novel ones like major customized computer systems (Morgan and Sayer, 1988, p. 24). The most obvious way to sell more products is to lower their prices, and lowering costs by reducing the amount of embeddedness in terms of interaction with customers is one way of doing that; thus, making computers more user-friendly not only makes them more attractive to customers but reduces the costs of supporting their sale.

Similarly, while there are different cultural forms of embedding, in competitive markets these succeed or fail according to how well they support profit. It has been widely noted that the Japanese form of embedding of economic processes differs from that of most Western societies, and it is tempting to interpret this as a fortunate cultural residue which has turned out to have unexpected economic benefits. However, there is some evidence that Japanese business had already introduced Western practices by the Second World War, and many observers argue that current Japanese forms of organization were actively constructed in the post-war period under the drive for economic recovery as a means of improving competitive performance (Eccleston,

1989; Fukutake, 1982; Dore, 1987). The latter organizational forms increase the competitive strength of Japanese firms by favouring the development of close, stable economic relationships, giving more emphasis to voice than exit, and favouring long-term development, innovation and quality improvement. These relationships are not based on sentiment or mere tradition but are subject to economic pressures and disciplines. Though deeply embedded relations between individuals and between enterprises are important in Japan, the relationships are far from being cosy: they can be conduits of relentless competitive pressure.

This is not to suppose that deeply embedded exchange relationships are *necessarily* economically advantageous: it all depends on what happens within them, on the norms, incentives and sanctions involved. Where the incentives for change are weak, deeply embedded relationships can be costly and induce inertia when radical change is needed. While markets are embedded in various ways, part of their dynamism derives from the threat of exit. Japanese firms may make greater use of voice, but the voice carries weight both because of the social and economic sanctions of a repressive collectivism, and because of the retention of the option of exit in the last resort.

Herein lies a major paradox, for despite being socially embedded, markets and money are also widely thought of as having a *dis*embedding effect, liberating individuals and practices from particular contexts. Money provides a form of valuation which abstracts from the particular use values of commodities, and the characteristics of the possessor. It is liquid – both indifferent to, and acceptable across, a vast swathe of contexts. If I have money, I can use it wherever it is legal tender, largely regardless of who I am, and I can spend it on anything (Simmel, 1990). This remarkable social indifference or neutrality can be seen both as a strength and a weakness. It gives economic relations an extraordinary flexibility compared to those of non-monetary systems, and is of historic importance in freeing individuals from traditional obligations and ties. On the other hand, this very indifference, and the opportunites for actors to take the exit option and switch to another buyer/supplier, invite the charge that money and markets foster social breakdown (Seabrook, 1990). The atomistic individual associated with liberal thought reflects this effect in a more positive light. Frequently one finds close combinations of disembedding tendencies and embedded behaviour; in the stock-market buying and selling takes place without reference to use values or the persons whose livelihoods may be disrupted by the transactions, but the traders and other market operators themselves belong to a particular culture, with its own procedures and norms, informal entry rules and patterns of socializing.

The apparent conflict between the emphasis on embeddedness and the recognition of the disembedding effects of money and markets is easily resolved. As Durkheim emphasized, money and markets do not allow people to escape from social relations *per se*, but situate them in different kinds of social relations from those outside markets (Durkheim, 1953). However, they do afford considerable freedom to agents to change the *particular set* of social relations within which they live (Kymlicka, 1990). A wage labourer is defined by social relations (in particular the relation with owners of the means of production), but she can change jobs and employers. The fact that individuals in capitalism are not isolated but socialized is perfectly compatible with the idea that they are relatively disembedded in the sense of being able to change or being obliged to change some of the social relations in which they live. In other words it is possible to acknowledge without contradiction both freedom of contract and the embedded and socialized character of market processes. Furthermore, there need be no contradiction between insisting on this embeddedness and in worrying about the corrosive effects of commoditization on social cohesion, since the latter prospect involves not the disappearance of social relations but a shift to a different kind of social relation which is selective and transient.

Contexts and power

What happens in a market exchange (or non-market exchange) depends a great deal on the circumstances in which individuals or institutions enter the relationship. The link with production is especially important. While markets are usually closely tied to production, the essential nature of market exchange is often explained by reference to examples of markets which are divorced from commodity production, or in which the connection is not specified. Thus the example is sometimes given of a prison camp in which the prisoners receive gifts from outside, which they then proceed to exchange among themselves. At the end of this process, each prisoner should be better off; indeed a Pareto optimal state should be reachable in which no one can become better off without someone else becoming worse off. But this is a rather peculiar market to use as a model of a market economy because the goods being exchanged were not sent to the market in order to be sold but were sent as gifts; hence the prices need not bear any relation to the costs of production of the items, and the original producers are not able to gain any revenue from this market. In this case, exchange is separate from the sphere of production and the division of labour.[5]

---

[5] Marx was clearly aware of how the incidental sale of products differs from routinized exchanges in commodity production (Marx, *Capital*, vol. 1, p. 166).

Markets are more normally intimately connected to production and the circulation of money or capital. Actors plan the supply of goods for sale and consumers – especially industrial ones – make their purchases in order to gain an advantage. How markets work depends on these connections to production and consumption. This can be demonstrated by reference to labour markets. Whereas the producers of say, soap powder, can gear its production to take most advantage of prevailing market conditions, the producers of people cannot. The weakness of wage labour derives not only from its lack of ownership of means of production but from the fact that people do not produce children with the future sale of their labour power in mind (and would presumably not do so even if they could anticipate future labour market conditions) (Offe, 1985). Labour power is consequently only a quasi-commodity. The combined effect of these circumstances is that wage labourers enter the labour market under duress, even though formally (if not practically) they are free not to do so.[6] This example illustrates both the significance of the relationship between production and markets and the more general point that despite the formal equality in which people or firms and institutions meet in the market, imbalances of power are common.

If we take a restricted view of a market, excluding production and consumption and the circumstances of buyers and sellers, then we are unlikely to appreciate these power imbalances. They are not only a matter of income differences, for actors also differ in their degree of organization, in their command over resources and information, in their possibilities for delaying transactions, and in the extent to which they are dependent on markets (Elson, 1988). If we are to understand what happens in markets we therefore need inclusive approaches that address such conditions rather than restricted ones which exclude them.

Markets are fields of power struggle in their own right, and prices themselves reflect the relative power of the participants in the market; they are 'estimated quantifications of relative chances in this struggle of interests', as Weber put it (1968, p. 108). Competition does not involve stable equilibrium, as the perfect competition model implies; rather it is rivalrous, with sellers trying to outsell each other, buyers often trying to outbid each other, and the interests of buyers and sellers at odds over prices. It is precisely these power struggles that lead to constant revisions in the costs or production and the (actual and perceived) use-value qualities of commodities (Lavoie, 1985). Yet the power differences do not result purely from success or failure in market competition, and even where they do, subsequent rounds of market exchange respond to the

---

[6] Though as we show later, workers may be better off as employees than working for themselves.

inherited distribution of assets, not merely to the current balance of *ex ante* demand and supply. The fortunes of actors at time *t* determine their market power at time *t*+1; someone who was (un)lucky last year has an (dis)advantage this year. Actors commonly engage in political manoeuvring to regulate and dominate both the market itself and its environment. Asymmetry is normal: Japanese firms as large as Toyota source some of their parts from two-person family firms; own-label suppliers of major supermarkets are often small and heavily dependent on a single customer, and while the supermarkets need these suppliers too, their survival does not depend on any particular supplier (Doel, 1994). At the same time, the balance of power is often difficult to estimate. Money is of course itself a major source of power, especially in capitalist economies, and its ownership is highly concentrated. Large firms therefore seem enormously powerful in relation to individual consumers. While this is certainly the case, as Offe and Heinze point out, whereas the supplier is stuck with the (illiquid) goods in particular markets, the buyer holds a liquid asset – money – which can be used to buy anything affordable (Offe and Heinze, 1992).

In some cases, entering markets can be hazardous to the point of threatening survival; the liberating qualities of money and markets can (contingently) lead to proletarianization and insecurity: to be 'free to choose' is to be 'free to lose' (Friedman, 1962; Roemer, 1988). For people who were previously largely self-sufficient, the switch to living by selling goods or labour power in markets may be enormously risky. For this reason Brenner has challenged the Smithian view of the transition from feudalism to capitalism (which many Marxists have accepted) as following spontaneously from the alleged propensity of individuals to 'truck, barter and exchange'. Against this he argues that the risks of this change from feudal and peasant economies to market economies were so great that it must have largely been precipitated by force, chiefly by expropriation (Brenner, 1986). Similarly, critics of 'structural adjustment' (i.e. marketization) strategies for developing countries, such as those of the World Bank, argue that, far from producing economic development, the normal workings of markets can greatly increase the insecurity of agricultural producers (Mackintosh, 1990; Olsen, 1993; Sen, 1981). It should be remembered that in some contexts marketization can be a euphemism for proletarianization. In Africa, British imperialists used devices such as the hut tax to force self-sufficient producers to enter commodity production and the labour market, so as to provide a labour force to exploit. In Thatcher's Britain, the creation of markets through privatization and liberalization of public services was used to shift power from public sector workers to the state and private capital. Marketization therefore both empowers and disempowers. In former state-socialist countries, the way in which production and distribution are marketized has enormous implications for the future distribution of power. Liberal economics is notoriously bad at recognizing these dangers of participating in markets. Its lack of interest in production and reproduction over time, and its treatment of exchange on the model of individuals freely exchanging goods and becoming better off in the process, mean that it fails to recognize the vulnerability of many producers.

References to the '*free*' market in neo-liberal discourse are so common, that it is tempting to think of markets as being zones of free, spontaneous conduct, unless interfered with in some way. But if markets are institutions, if they rely upon the state, if they are organized, regulated, embedded and influenced by context, if information and power are unevenly distributed within them, and if they can disempower as well as empower people, how can they be 'free'? Here, some careful balancing of arguments is needed. The fact that most markets are regulated in some way does not wholly undermine standard arguments about the properties of 'free' markets. In particular, it does not necessarily mean there is no choice. Most markets are only regulated on the supply side, and buyers still have the freedom to move between them. Only where there is no scope for choice and substitutes, as in the case of the British coal and nuclear-power markets, where major electricity generators are compelled by the government to buy certain quotas, can markets be said to be not merely regulated but rigged. At the same time, just because unrigged markets afford some choice to economic actors it does not mean that they are a free, unorganized space lying between the organized spaces of enterprises.

Competition

Market competition operates not only by firms acting as price-takers and minimizing costs according to known procedures (weak competition), but by trying out new procedures and products whose costs are not knowable in advance since they change the whole basis of pricing and create disequilibrium (strong competition) (Storper and Walker, 1989). Contrary to the impression given by mainstream economic theory, with its static-equilibrium approach and abstraction from technical and organizational change, it is strong competition which is important for economic development. Intermediate between strong and weak competition is strategic action, in which actors can roughly anticipate the impact of their actions on others (Baland and Platteau, 1993) and mould market forces instead of being conditioned by them (Lazonick, 1991). Where there are potential economies of scale and learning-curve economies – as in semiconductor production, for example

– innovating firms may drive down prices not because sales are too low but in order to build up sales so as to realize these economies and gain a competitive advantage over slower-moving rivals. Although such strategies are still constrained by consumer response, it is the dynamics of production and not simply market conditions which determine price. Nevertheless, we should not flip over into assuming that markets are entirely manipulated, and that the invisible hand has been stayed, as writers such as Lazonick imply. Markets can rarely be controlled completely; even firms with as much monopoly power as IBM had prior to the mid-1980s, and with the ability to lock in customers, can fail to anticipate the trajectory of technology, competition and markets. Strategic competitive behaviour does not displace the need for flexibility and opportunism. While firms can set as high (or low) a price as they like, whether they can realize sufficient sales is quite another matter. The development of business organization in production and distribution both shapes markets and is shaped by them. It can hardly be denied that competitive markets induce enterprises to lower costs and introduce new products, especially under capitalism. At the same time as competition encourages parsimony and efficiency in productive consumption of particular resources within enterprises, it also encourages them to defend their position by adding value to products and increasing total sales, with the result that their own consumption and that of final consumers expands inexorably.[7]

Behaviour in markets

What happens in concrete markets depends on more than the basic structure, for example in terms of the number of suppliers. It depends on what actors actually do. Structure constrains and enables, but it does not eliminate variety. Thus, markets with small numbers of competitors are not necessarily less competitive than ones with large numbers, for competition depends on the strategies chosen by enterprises, and these can vary independently of the number of rivals. While there are constraints on what actors can do in markets (e.g. the prohibition of theft), there is often considerable scope for variation in behaviour. Although, at an abstract level, all markets as we have defined them have certain essential features, in concrete cases they can be developed in various ways and combined with other modes of coordination so that

---

[7] This economizing is of course made in terms of monetary values which take inadequate account of the depletion of non-renewable resources and energy. For some 'deep green' critics, the contribution of markets to the growth imperative is sufficient to condemn them (Dryzek, 1987; see also Altvater, 1993).

behaviour and outcomes can vary significantly. In concrete markets, outcomes depend on the nature of products and technologies, on the forms of calculation actors use, on the legal and moral framework, and on the strategic knowledge of actors (Auerbach, 1988).[8] Nevertheless, variation in behaviour is not entirely random but patterned by economic pressures of competition and hard budgets and similarities in needs.

Markets and space

Markets require communication and the circulation of information between buyers and sellers, often coupled with a certain amount of trust-building. In most cases goods have to be physically (sometimes electronically) delivered to the buyer, the exception being those, like shares or currency dealing, where purchases need only to be recorded, not physically appropriated. Since travel and communication are costly, markets are *spatially differentiated*. We often talk of national or even world markets, but to sell in them one generally has to be able both to communicate with customers in particular places and deliver goods to them, and sometimes provide after-sales service too. We might say that a British firm has an $x\%$ share of the world market for a certain product, but often it is only active in a few national markets. To get into more it usually has to have a physical presence in them. Global communications are rarely sufficient for this purpose. Hence a British electronics firm aiming to sell products in the USA, may itself be a sophisticated user of such technology, but it will generally have to locate a visible and credible sales operation (perhaps in concert with other operations such as R&D), close to customers such as telecommunications firms, if it is to stand any chance of success (Morgan and Sayer, 1988). Such markets are in this sense highly localized, even though transactions within them are generally considered to take place within a national or world market. Similarly, Chilean onions are not in the same market as English onions unless they are brought to the same places, where they can be sold to the same customers. Though many commodities are sold all over the world, the world market for a product is therefore usually the sum of all the

---

[8] Tomlinson (1990, p. 43) argues that 'There is no such thing as "the market", but only markets, whose effects are determined by the agents who operate in them, their forms of calculation, their relations to other agents and the legal and moral . . . framework in which they operate' (see also Auerbach, 1988). Thus, those whose job is buying or selling for a company will behave in different ways from the amateur individual buying for his or her own consumption. As we argued in chapter 2, this is a valuable point regarding concrete behaviour, but it does not mean that anything can happen, so that markets have no abstract qualities or essences which abstract political economy can theorize.

national or regional or local markets rather than a single, continuous market. There may be potential buyers in most countries for products like grain or crude oil, but the market institutions and suppliers to which they must go to buy it are not ubiquitous but localized, though we commonly abstract from these spatial specificities in talking about such markets.

Physical space, and the costs of overcoming it, therefore give markets a degree of spatial monopoly, as central-place theory demonstrates (Berry, 1967). Trying to create spatial monopolies for one's own products, cornering a particular group of customers all to oneself, and trying to attack the spatial monopolies of others, are key kinds of competition. Thus, precisely because the costs of establishing markets – of bringing together sellers, products, information and customers – vary across geographical space, market location strategies are often vital to competitive success; witness the store location strategies of supermarkets. On the other hand, to the extent that the boundaries of spatially specific markets are often under challenge from possible invaders with cheaper or superior goods from outside, it is understandable why we should so often think in terms of single aspatial or pervasive markets. Suppliers, goods and customers can often move around so that the boundaries are flexible. We might say that they can move between markets, but then they also help to constitute markets themselves. The spatial scope of markets is consequently often ambiguous or multi-levelled: Do markets for, say, sportswear exist at the level of individual urban centres, countries or globally? Is a supermarket or department store a market in itself? We might talk of the world market for plutonium, but effectively it is limited to a few special places. The ambiguity is deepened further by a common slide between our root sense of markets and another sense of 'market' as *potential or actual demand for some commodity*, as in 'there's a big market for insurance in Russia'. References to a 'national' or 'global market' usually play upon this latter sense. In this usage of 'market', there may not be any fuctioning markets in our root sense in existence yet, but merely some of their preconditions.

The Multiple Meanings of 'Market'

Such is the variety of uses of the term 'market', that it is reasonable to talk of multiple concepts of markets. As Maureen Mackintosh observes, these are rarely distinguished, so that people regularly slide unknowingly between quite different uses of the term, sometimes within one sentence, while imagining that they are talking about the same thing (Mackintosh, 1990; see also White, 1993). These conceptual slides are a feature of both lay and academic/scientific usages of the term, and are found in both right-wing and left-wing, liberal and radical economic theory. The Right tends to pass off idealized models of markets as descriptions of actual markets. One of the first things mainstream economics tries to do in inducting new students is to encourage a rationalist perspective in which questions of the adequacy of abstractions are not addressed; in particular, students learn to slide between different senses of 'market' without worrying, as if conceptual precision was worth sacrificing for the sake of generality or making models analytically soluble. The Left often responds to liberal uses not by deconstructing them, but by merely reproducing them in more negative tones, so that markets are associated with any and all the ills of capitalism. In everyday usages the shifts are often innocuous. Polysemy is not necessarily a problem. The scope and subtleties of everyday usages are certainly worth respecting and may even be something we should admire. By comparison, attempts to analyse and codify the tacit understandings involved are bound to seem lumbering and unsubtle. Nevertheless, the ideological influence of the discourse of markets is too important for the conceptual slides to be ignored. While the variety of different uses of 'market' and 'markets' is confusing, many of the uses have contexts in which they identify something significant. The problems come when authors apply them outside these contexts, particulary where explanatory weight is transferred unknowingly from one referent of 'market' to another. On occasion, the conceptual confusion can have disastrous effects; Mackintosh found a World Bank report offering diagnoses and prescriptions for poor countries to have 'at least three different meanings floating in the text.' (Mackintosh, 1990).[9] In this section I shall attempt to take further her strategy of distinguishing different senses.

I shall argue that concepts of markets differ according to:

- their level of abstraction;
- their inclusiveness;
- whether they are couched within a 'market optic' or a 'production optic';
- whether they refer to real or imaginary markets.

I shall deal with these in turn. As is often the case in this kind of conceptual analysis, it is difficult to stop using the very concepts one is criticizing; nevertheless, I shall try!

[9] These are: 'the market', denoting exchange of goods and services, including labour power, for money, and private ownership; secondly, abstract models of markets constructed by economists; and thirdly, different ways of buying or selling i.e. the concrete or real markets studied mostly by anthropologists and geographers.

Levels of abstraction

Real markets may be conceptualized at different levels of abstraction. Thus, one can talk about the local fruit and vegetable market relatively concretely, in terms of who the sellers and buyers are, how it is organized, the labour involved in running it, how it is regulated, how actors compete and cooperate, what exactly they sell, how they are constrained by their situation (e.g. access to credit, ability to wait for conditions to change). Alternatively, we can look at it more abstractly, in terms of the exchange of commodities and property rights for money under competitive conditions, or still more abstractly as a mode of coordination of the division of labour.

Formalized market models, such as the perfect competition model of neoclassical economies, are highly abstract, though they tend to be designed not so much as representations of actual markets as devices for building deductive, and thus analytically tractable, puzzles. The models are of course dependent on a raft of assumptions, some of them harmlessly unrealistic, others negating circumstances which are not only universal but make a major difference, such that relaxing them would undermine the logic of the model. The latter kind of assumption is harmless when the models are treated merely as puzzles, but seriously misleading when applied. Assumptions which merely simplify and approximate or hold off contingently related variables are one thing; assumptions which negate the essential and causal features of real markets are another (e.g. the treatment of time and knowlege in the perfect-competition model) (Sayer, 1992a).

Inclusiveness

Concepts of markets differ not only in level of abstraction but also in inclusiveness. Markets may be defined narrowly in terms of routinized buying and selling under competitive conditions, or inclusively to embrace not only exchange but the production and consumption of the exchanged goods, and the particular property relations that hold therein.[10] Accordingly, for the fruit and vegetable market we could adopt a restricted focus, ignoring what buyers and sellers do outside of the moment of exchange, or we could take a more inclusive view, examining how the sellers get their produce, how the supply chain is organized, even going back into production, and on the other side, how customers are differentiated into individuals and institutional buyers and how their

---

[10] Devine (1988) suggests a distinction between markets and market forces, corresponding to our restricted and inclusive concepts.

purchasing behaviour is related to things like income and ability to save. In both cases, we could develop more or less abstract accounts of these. It may thus help to picture abstraction in the vertical dimension as a matter of 'levels', while inclusiveness is pictured in the horizontal as a matter of breadth.[11]

For a market or commodity-producing economy it is probably unwise to try to draw a boundary between the market, or the sphere of exchange, and production and consumption, since the development of markets may only be intelligible in the light of developments in production and vice versa. Where markets involve the exchange of goods which were not intended for sale when produced or which are being re-sold after a period of use (e.g. antiques), the market has more autonomy from production, and a restricted analysis of the market may be adequate.

Many abstract discussions of 'the market' slide between restricted and inclusive versions. The basis for the powers or forms of behaviour commonly attributed to markets are consequently often ambiguous; is it markets in the restricted sense which give rise to the effects of interest, or markets when mediating between particular kinds of producers, with certain kinds of property relations, and particular kinds of consumers? As we shall see, markets can coexist with different property relations and systems of production, so we cannot expect to read off an inclusive account from a restricted focus. This is of critical importance in political economy for understanding and evaluating market economies. Restricted accounts of markets exclude major contextual influences which explain behaviour. On the other hand, a more inclusive view which takes in those influences is going beyond the market into production and consumption. The dynamism of capitalist economies is not simply a consequence of markets in the restricted sense, but of capital, obliged to accumulate in order to survive, and liberated from the ties which bind petty commodity producers.[12]

Marshallian demand – supply diagrams provide restricted views of markets, marginalizing the social relations of production and the processes of production and consumption on which demand and supply depend. As Maurice Dobb pointed out in 1937, this treats production and consumption as the creature of price rather than vice versa. Moreover, the static-equilibrium approach, with its conflation of *ex ante* and *ex post* quantities, treats markets as closed systems. Instead of tracing the aetiology of actual markets in an inclusive sense, taking into

---

[11] See Mäki, 1994, for more detailed analyses of abstraction.
[12] Ideally it would be useful to know what market behaviour is like under different production relations. In principle, a disaggregative approach has the merit of trying to address this problem, though it is extremely difficult to answer in practice.

account the semi-autonomous dynamics of production and consumption, this approach attributes change either to an exogenous black box (technologies and preferences) or to the endogenous variable of price signals (Dobb, 1937).

Production and market 'optics'

The problems regarding inclusiveness are frequently combined with those of broader conceptual frameworks. In practice, decisions about what is included are largely determined on the Left by a 'production optic', in which production and capital are generally treated as prior to exchange, thereby marginalizing markets, and on the Right by a 'market optic' which swallows up production in exchange. In the former, commodities are merely 'thrown' on the market, and the role of exchange is limited to realizing the value of commodities and completing the circuit of capital. Allocational effects are of no interest. By contrast, the market optic focuses on the allocational effects and either ignores production and its social relations, or conceives of it as a sphere of transactions or exchanges, or else reduces it to 'supply' (which fails to distinguish between produced goods and unproduced goods such as land). In the market optic of mainstream economics, the whole economy becomes 'the market' in the singular, or 'market system'. In economics textbooks, markets are commonly counterposed to the state, as if markets – or 'the market' – were everywhere the state was not.[13] Yet individuals and indeed many organizations, actually spend rather little of their time making transactions in markets. People going about their own business, outside of bureaucratic or state control, can do a lot of things which don't involve buying and selling even if they also intermittently depend on it. Thus, when I am writing this book I am outside both state and market, even though they are a condition of my way of life.

One of the most extraordinary effects of the market optic is the illusion that markets actually produce things. This is evident in the neo-liberal rhetoric of the market both in the West and in former communist countries where markets are expected to 'deliver the goods'. Arguably markets in the restricted sense may be good at stimulating production, but certainly they produce nothing themselves. The illusion is further evident in the assertion that firms can either make things or get them from the market (the 'make or buy' decision). In fact, of course, the existence of a market for inputs is dependent on the existence of other producers. Markets are not an alternative to production or to firms or

[13] 'Civil society' is therefore hidden in this conceptualization.

'hierarchies', but a mode of coordination of the division of labour. Enterprises or hierarchies are usually involved not only in coordination but production; they are therefore not merely an alternative mode of coordination, as mainstream economics tends to assume.

It could be objected that it is naïve to take the idea of markets producing things as anything more than a shorthand: no one seriously imagines that exchanging things actually produces anything and it is *obviously* the whole 'market system' which is meant here, including producers – or rather, 'entrepreneurs'. Doesn't everyone know that buying from the market is just a shorthand for buying from other producers in the market? However, the concept of 'the market' or 'market system' here is not merely an inclusive one which encompasses production, for as we have seen, the market optic barely acknowledges production, reducing it to 'supply'. The shorthand would not be suspect if liberal economists took production seriously and if the belief that exchange, or change in ownership, actually created anything new were not so common in actual economic behaviour. The illusion is particularly strong in the more liberalized capitalist economies, where it is evident in the pursuit and celebration of takeovers. In such cases we have markets without production – the buying and selling of existing commodities or property, including companies, without adding anything to the existing stock of goods and services. Treating this as equivalent to productive activity is a disastrous error, though it might facilitate – or inhibit – *subsequent* production, and *might* affect efficiency. The confusion is echoed further in the Thatcherite practice of encouraging share owner- ship – i.e. rentiers – as a way of promoting entrepreneurship: whereas entrepreneurs are celebrated precisely for producing something which didn't already exist, rentiers can earn an income without producing anything. There are important arguments about the benefits of market exchange in terms of their effect on production and the efficient allocation of resources, but it remains the case that exchange itself produces nothing. There is a curious fetishization of the power of markets here, which attributes to a mode of coordination, a power which is actually dependent on another sphere of the economy. It is the liberal celebrants of 'the market' who are fooled by their own shorthand.

Further grounds for concern over the difference between inclusive and restricted analyses of markets relate to the ideological uses which can be made by slides between the two or attempts to pass off one as the other. For liberal economics, the market optic and the use of a restricted analysis of markets as if it were actually an inclusive analysis, marginal- izes social relations and presents capitalism in neutral guise, merely as a market economy, as if it were not significantly different from non- capitalist market economies of petty commodity producers. In Marxism,

matters are reversed, with the behaviour of market forces under capitalism being taken as condemning markets operating outside capitalism.

Metaphorical, imaginary and latent markets

We can further distinguish literal concepts referring to real markets from
those referring to imaginary, latent or implicit markets, and from those
using market metaphors to refer to real situations which have only
limited similarities with real markets, for example ones which do not
involve any transactions of commodities and property rights.[14] The
metaphorical uses involve abstractions, but it is not the level of abstraction so much as their quality which is important; 'the marriage market',
for instance, is at best a fanciful metaphor in most Western cultures, as
marriage partners are not actually bought and sold for money, even
though some may have more success in finding the partner they want
according to their beauty, wealth, etc. (Hodgson, 1988).

Concepts of imaginary, latent or implicit markets figure prominently
in mainstream economics. It should be noted that these are not the same
as abstract (i.e. one-sided, selective) concepts of a real market (in which
commodities and property rights are traded), for whereas the latter
already exists, imaginary markets are only a possibility. Hodgson (1988,
p. 81) notes how Arrow and Debreu assume 'that a market exists for the
exchange of every possible commodity on every possible date in every
possible state of nature'. This is an extraordinary usage, for it means
almost the opposite of what it says: namely that such a market hardly
exists and can only be imagined. What this concept of market seems to
involve is a representation of economies as consisting of a vast array of
opportunity costs, where the goods whose use or non-use has these
opportunity costs could be exchanged for money in markets. Once
everything is seen as having a price, notional or real, then it is tempting
to look upon the range of resources and projects in society as one big
market. A loaf of bread, a picture, a house, a field, a letter, a haircut, a
motorway, a worker, a conversation – all these things and countless
others might be thought of as having a price which someone might
possibly be prepared to pay for them, though even under capitalism,
some of them may never be offered for sale. In some economies few or
no goods are exchanged in real markets at all, and as many have pointed
out, including Marx, it is absurdly ahistorical (and often ethnocentric)
to project market concepts onto non-market economies, and therefore

---

[14] There is no objection in principle to metaphors in science, indeed they are indispensable,
but there are some which are fanciful rather than the best available representation of
their objects.

quite accurate to term this kind of economics 'bourgeois'. The assumption that anything and everything can and should have a price, and be tradable, though influential in capitalist society and endorsed in liberal and libertarian theory, is questionable (e.g. Sagoff, 1988). All societies exclude at least some things, whether privately or communally used, from being traded, and non-Marxist as well as Marxist critics of markets often argue that the market sphere is too large in capitalist societies (Walzer, 1985).

However, while Marxism's critique of this way of thinking is still powerful, it remains the case that it does, despite its unhelpful terminology, contain an insight, which Marxism has great difficulty in understanding, namely that even if there are no real markets, there is an array of opportunity costs regarding the use of resources, including labour power. Even in the most abundant environment, in pre-industrial societies where food gathering and other labour takes little time, there are always opportunity costs or different degrees of allocational efficiency. Things could always have been done in different ways, with different trade-offs and costs and benefits, even if they are invariably done on the basis of convention rather than calculation: the absence of markets and prices or any form of economic calculation does not mean the absence of opportunity costs, and conventions are not necessarily inefficient. That there are always opportunity costs is a non-trivial, transhistorical fact about all economies. This is a consequence of both constraints and potentials or opportunities (there are only 24 hours in a day, individuals can't do many things at once and be in several places at once, resources have limited possibilities for use and are subject to the second law of thermodynamics). In simple economies, in so far as opportunity costs are estimated and compared, they are fairly transparent and may be evaluated in real terms (i.e. in terms of use values) without the aid of money and prices. In advanced economies they can be evaluated indirectly and abstractedly in money terms. *The conception of an economy as an array of opportunity costs is a useful one, but since the existence of such an array is only contingently related to real markets, it is silly to refer to it as 'the market'.* Real markets are just one form in which those opportunity costs sometimes get reflected.

We could say that this concept refers to 'implicit markets', but this still suggests that real markets are the normal form of economic organization, and if absent, are held back by pre-modern conventions and practices, ill-defined property rights and state restrictions, and just waiting to be 'freed', whereupon economic benefits are supposed to follow. In this way, the conceptual slide from imaginary to actual markets is closely associated with negative judgements of non-market production and modes of coordination as causes of economic backwardness, and it has the effect of legitimizing policies for the development of actual markets. Thus commodity production is assumed by the World Bank to be superior to subsistence production and state regulation, despite the plentiful evidence that marketization in developing countries offers no guarantees of development (Mackintosh, 1990; Sen, 1981) In liberal economics, 'the market' is privileged both normatively as the best form of economic organization and positively as the key to how actual economies work; indeed the image of the former colours its vision of the latter. Economies which lack markets or only have few are judged negatively, not only because they lack mechanisms which allegedly bring benefits, but because they don't fit with mainstream economics' market optic. Further, 'the market' or 'markets' are given powers and authority of their own and treated as if they were unitary actors. In a sense markets do condense all the demands and offers made, not only in a single market but in others in which the same actors participate, and across which resources are allocated. Yet as we noted earlier, market prices are not merely neutral reflections of demand and supply but reflect the balances of power in many arenas. 'You cannot buck the market', a slogan beloved of Margaret Thatcher, was another way of saying that 'might is right', regardless of how the might was distributed.

The ideological notion of latent or implicit markets which only need freeing figures strongly in neo-liberal rhetoric, and contrasts strikingly with the view, associated with Polanyi and others, that markets are social constructions whose birth is difficult and requires considerable regulation and involvement by the state and other institutions to achieve (Polanyi, 1944; Marquand, 1988). The experience of the post-communist countries weighs heavily in support of Polanyi. The liberal underestimation or denial of this institutional support is partly derived from the elision of the difference between potential or imaginary and the actual in its concept of 'the market.' In this kind of view, 'The market is seen as an ether in which individual and subjective preferences relate to each other, leading to the physical exchange of goods and services' (Hodgson, 1988, p.p. 177–8). The ether is the latent or imaginary market, and again actual markets are seen as their natural consequence, unless somehow blocked.

While the liberal tendency to see markets everywhere is supported by this concept of imaginary or implicit markets, Hodgson (1988) suggests another source, one which derives from seeing all social actions as *exchanges* between atomistic individuals, and seeing contracts as the model of social relations, ignoring intersubjective rules and norms. That only some of our actions involve choice, and then in terms of social relations, intersubjective rules, etc., is not commensurate with the rationalist methodology of liberal economics. Thus the interactions between lovers or parents and children, are seen as exchanges. Here the market optic extends beyond the economy. The exchange model of social action is also congenial to those who want to make individual 'choice' the organizing principle of economic behaviour, rather than production or social organization. For example, J. Buchanan defines the market as an institutional process 'within which individuals interact, one with another, in pursuit of their separate individual objectives, whatever these may be' (cited in Brown and Harrison, 1978, p. 87). In so far as this need not involve the exchange of money for commodities and the exchange of property rights, the 'market' here is imaginary and metaphorical.

Economic discourse – including radical political economy – is plagued by elisions among these different concepts of market. Uses of 'the market' in the singular are particularly slippery, referring either to a specific real market, or to the whole system of such markets throughout the economy, or sometimes to the allegedly latent 'market' discussed above. Notwithstanding the tendency for 'the market' to conjure up a picture of a society of equals, freely entering into contracts, and in which production and its social relations are nowhere to be seen, the second of these three uses – the system of real markets – has some logic in that markets are interdependent, such that changes in a particular market (e.g. the oil market or the mortgage market) have effects which ripple through entire economies, indeed round the world. This notion of 'the market' in the singular, as a pervasive system of particular, interlinked markets, fits better with the perspective of the buyer or investor, having money and able to spend it on anything, than the seller who is stuck in a particular market with the particular commodities he or she has to sell (Offe and Heinze, 1992). The holder of money can roam across many markets, as if they were one big market. Once again, while the many meanings of markets can cause considerable confusion, there are often contexts in which particular uses contain a valuable insight. In developing a critique of the discourse of markets we have to recognize these insights as well as attack the conceptual elisions.

Other Modes of Coordination

Contrary to the impression given by liberal economics, markets are not a self-sufficient mode of coordination, but invariably need supplementing by other modes too. These other modes are also rivals to markets, though they too cannot coordinate advanced economies alone; the different material and informational coordination problems thrown up by concrete economies are inevitably too diverse for this and require hybrid solutions (Hodgson, 1988). However, once again, we can form a clearer understanding of concrete situations on the basis of prior abstract analyses of their constituents. I shall briefly outline same of the main features of planning, networking and democratic control. (For fuller accounts, see Lindblom, 1977; Dryzek, 1987; Sayer and Walker, 1992).

Planning

Many segments of the division of labour are directly coordinated a priori by planning and bureaucratic control. In commodity-producing economies, the operations involved in producing a particular commodity must be planned, be it a one-person firm or a multinational. Markets can therefore never be a complete alternative to planning and hierarchy, and certainly not to production, as implied in the misleading markets-versus-hierarchy dualism popularized by Williamson (1975, 1985). Bureaucratic control, in Weber's non-pejorative sense, is the norm for organizations of any scale, whether operating in markets or outside. It is by no means exclusively associated with public ownership; private organizations need a significant degree of bureaucratization if they are to cope with large throughputs of information and materials. Though bureaucracy has well-known deficiencies, especially with respect to flexibility and motivation, it is efficient, and even the most post-Fordist of firms need significant degrees of hierarchization and routinization of activities in order to function.

Planning must be authoritarian and centralized to a significant degree. Democratic referring may be possible, but design and implementation must override local or sectional opposition, or the plan will be ineffective. This applies both to planning within organizations and across the whole economy. But as Hayek's economy*/catallaxy distinction makes clear, and as socialists often forget (e.g. Altvater, 1993, p. 25), planning a technical division of labour within an enterprise or institution is not equivalent to planning a whole social division of labour, as in central planning in state socialism. There is a world of difference between planning how to make a single commodity like a telephone or shirt and planning the allocation of resources across the production of millions of different commodities aimed at millions of consumers with different and particular needs. Both require centralization but in the former case the organization has coherent, complementary, objectives, whereas in the latter it does not. Central planning of a catallaxy is not a matter of finding a solution to a particular problem or a means towards a particular end. It does not involve solving a set of equations containing details of user preferences and available resources, so that the economy's vast input–output matrix can be coordinated and an equilibrium reached, for this information is not obtainable (Nove, 1983). The required data only exist in neoclassical models of economies, not in real economies. In real market economies, strong competition and the largely unpredictable price shifts it stimulates provide a discovery mechanism, allowing comparisons of the costs of different allocations of resources to be made. Preferences are not fixed prior to and independently of supply and price, but develop partly in response to whatever is available (Buchanan and Vanberg, 1991). Without these mechanisms there is a fundamental problem of coping with new, unforeseen needs and difficulties, and it is this, rather than the neoclassical problem of finding an equilibrium, which presents the biggest obstacle for socialist calculation under central planning. Here the *ex post*, decentralized control afforded by markets, coupled with rivalrous, strong competition has an advantage over *ex ante* control through planning (Lavoie, 1985).[15]

Inevitably, in practice, the planners have to decentralize decisions to a significant degree, and to aggregate and approximate needs and production capabilities. This produces inefficiencies in micro-economic coordination – just what kinds of electronic components, or drugs, are needed? Millions of users cannot value millions of products by voting and central planners cannot value them (in terms of use value) on their behalf, except in the crudest terms; exchange value and money are not eliminated in advanced, centrally-planned economies. The state must therefore either decide what people should consume, and allocate it as use values, or allow them to 'vote' with money by buying commodities. The latter course predominates because it is so difficult to register relative scarcities and costs when the economy is coordinated through use-value targets. Hence exchange value is retained, though prices are not market prices and do not influence the economy in the same way (Arnold, 1989; Brus and Laski, 1989; Lavoie, 1985). One does not have to idealize market prices to recognize that these regulated prices were unlikely to reflect relative scarcities and ration consumption and production accordingly. They produced distortions and disequilibrium on the supply side and encouraged wasteful consumption; for example, in the USSR bread was so cheap that it was widely used as animal fodder. Fixing prices purely according to costs ignores use values, demand and opportunity costs. The average socially necessary labour time for producing a pair of shoes or a pair of skis tells us nothing about the relative use values or social necessity of shoes and skis. On the other hand, making relative

---

[15] Those socialists who have defended the possibility of planning along the lines of the Lange–Taylor model can meet neoclassical objections because they share the same premises regarding information and restriction of competition to the weak variety. What they cannot do is meet these Austrian objections (see also Steele, 1992).

need the sole criterion ignores questions of allocational efficiency in meeting those needs. Lastly, setting plan targets in output terms rewards waste compared to allowing demand to regulate output (Nove, 1983). Therefore, as Weber and many liberal critics have argued, centrally planned economies lack formal rationality (Weber, 1947, pp. 148, 243)

In practice, central planning tended to be heavily influenced by political considerations, often at the expense of economic efficiency. Leante (1980) claims that in this respect it combines the worst of centralism and localism. One of the paradoxes of centrally planned economies was that the very impossibility of comprehensive planning made them susceptible to influence by powerful local or sectional interests, such as that of heavy industry in the former Soviet Union. Conflicts between interests of the centre and particular production units were inevitable – it being often in the latter's interest to deceive the centre in order to protect themselves. Although the bureaucracy controlled and distributed the surplus, enterprises were far from powerless and could withold information from the centre and leverage permissive funding. The incentive to enterprises to exaggerate their needs in order to protect themselves in meeting targets was a major cause of shortages. Not surprisingly enterprises developed contradictory 'adolescent' relationships to the centre, both wanting more autonomy and demanding more paternalistic favours. At the same time they bartered and made deals with suppliers and user-enterprises through informal political networks. In some periods the inadequacies of central planning were so great that these practices became the dominant mode of micro-economic coordination (Kornai, 1986; Burawoy and Kritov, 1993).

In no other situation is the complexity of the social division of labour and its nature as a catallaxy so apparent; ironically, the attempt to abolish or suppress the social division of labour through central planning only makes its materiality clearer. In contrast to the hidden workings of the market mechanism, centralized allocation of resources among competing ends is transparent and highly politicized. It is not so much the class, bureaucratic or undemocratic character of central planning which causes its distortions and inefficiencies, but the enormity of the task of planning a complex economy which makes bureaucratic structures and authoritarian decision-making inevitable (Nove, 1983). Not surprisingly, in practice, central planning had to be patched up and supported by a host of informal deals between enterprises and institutions to work at all. However, looser forms of planning may be more feasible and compatible with markets.[16] In 'indicative planning' attempts are made to

---

[16] Hayek's insistence that any 'interference' with markets necessarily leads towards full, centrally planned socialism has been widely refuted (Tomlinson, 1900b; Wooton, 1945).

coordinate independent producers. The latter may find it useful to comply with such plans as they may stabilize expectations, overcome assurance problems and reduce risks and the problems of regulation endemic in market economies, but by definition, they cannot be forced to comply.

Networking

Even in a highly advanced social division of labour there are areas in which closely related activities can be coordinated directly through various blends of collaboration, negotiation and domination. These usually involve small numbers of actors with common interests, well-defined projects and the possibility of direct interaction. Networks consist of interlinked units exchanging information and services for mutual benefit. They are particularly useful for building trust, which is a precondition of any successful ongoing economic relationship. They can be either hierarchical or non-hierarchical, and they can exist both within and between organizations. Networks share external economies and treat information as a public good. As we saw in the last chapter in the discussion of Hayek's concept of the division of knowledge, while each enterprise or actor may have unique, specialist knowledge and information – at least about their own needs – significant amounts of information are common to and required by groups of actors or enterprises. Contrary to the fiction of markets as a self-contained mode of coordination needing only price information, actors cannot rely on merely 'going to the market' to see what is on offer and to see who will buy their wares; groups with common interests – especially producers – need to share information, negotiate, form common views about the future and harmonize plans.

According to recent studies of industrial organization concerning 'post-Fordism' and 'flexible specialization', networks, in the guise of alliances, industrial groups, relational contracting, etc., are gaining a new significance as a mode of coordination of flexible production (Best, 1990; Piore and Sabel, 1984; Murray, 1989; Scott and Cooke, 1988; Sabel, 1989; Sayer and Walker, 1992). While these developments have generated extensive debates, one point is worthy of note here. It concerns the impression given by the literature that networks are beginning to supplant market relations and the price mechanism. They may be supplanting arm's-length market transactions to a certain extent, but the fact that the members of the network remain under separate ownership and still exchange most of their products as commodities, shows that networks usually supplement rather than displace market exchange. They modify and regulate markets, but do so over restricted bundles of activities within production chains. Exchange value remains the bottom line for companies. Although, as the literature shows, actors can interpret market signals in different ways, not just any interpretation will be successful. If a firm's products suddenly stop selling, no amount of re-interpretation will change that fact in the short run. We must therefore beware of an implicit 'oversocialization' in the representation of economic relationships in the post-Fordist literature. The ambiguity of the popular phrase 'beyond markets and hierarchies' should be noted: it refers to the supplementation, not the displacement, of markets and hierarchies by other forms of coordination (Sayer and Walker, 1992). Finally, it must be remembered that this literature focuses primarily on particular industries and their internal supply chains; but the price mechanism is also largely responsible for governing the allocation of resources between quite different industries within a catallaxy rather than an economy*, a matter of great importance not touched by this research. Unlike markets, networks can only function with small numbers of members and in relation to relatively simple and interrelated goals.

While networks involve sharing and collaboration, we should be wary of assuming that the members of the network are necessarily equal in power, particularly when the network supplements market relations. Network relations need not be 'cosy' and democratic. It is often easier for a large firm to form a network with subordinate suppliers than it is with firms which are equal to it in power, as equals may have less need to give up information in order to get business. Clearly, sharing information with competitors is also less likely than sharing information with suppliers and customers, but even the latter involves an element of competition for added value within the supply chain. In such situations the dominant firm can share information where this suits its interests and still dominate its suppliers, who may have little option but to comply. It is not in the dominant firm's interests to drive its suppliers to the wall but it can still drive a hard bargain. Information useful for collaboration is also useful for surveillance. This is cooperation on the parent – child model rather than cooperation among equals. In any competitive system, or in an organization facing competitive pressures – and it need not be competition in a market – the challenge is to share information where this is beneficial, to withhold it where it endangers self-interest, and to win cooperation without losing discipline. Competitive situations often generate problems where individual agents are deterred from contributing mutually beneficial information for fear that others will fail to reciprocate and free-ride. It is therefore important to avoid romanticizing network coordination. Networks are supposed to reduce these problems but they often need the assistance of a body standing outside the competitors, such as the state, an industry association or a council to override these individualistic pressures. It bears repeating that this applies to any competition, not merely market competition.

Democratic control

Although not generally thought of as a mode of coordination, democratic processes can, to a certain extent, be used for this purpose. Unlike other modes, democratic control is not just a means to an end but has intrinsic virtues as an end in itself, though these may not compensate for its defects. Its scope is restricted by the size of the constituency, and the number and complexity of issues at hand. While democratic control can be made less costly by delegation and representative democracy, the costs may still outweigh the benefits (Buchanan, 1985). In addition a free-rider problem weakens the incentive to vote (my vote won't make any difference). Electoral democracy is limited to voting either for simple single issues or for whole packages of issues – disaggregation is extremely costly. Where interests are diverse and the tasks complex, the degree of control can only be loose, not least because of the division of knowledge. Millions of users cannot vote for millions of different products from thousands of different enterprises. Nor can 'an elected assembly decide by 115 votes to 73 where to allocate ten tonnes of leather, or whether to produce another 100 tons of sulphuric acid' (Nove, 1983, p. 77). Socialists' enthusiasm for democratic control of a whole economy depends on ignorance of these practicalities.

It is in the very nature of an advanced economy that society comes to depend on arcane bodies of specialist knowlege which are largely beyond the understanding of even highly educated outsiders. How many voters can be expected to understand the options open to the Chancellor of the Exchequer? This also presents problems of accountability for any mode of coordination: how can central planners assess whether the specialists' bids for resources are reasonable?; how can consumers know whether the providers of a specialist service like medicine can be trusted? As in economics, so in politics: the scope for effective democratic control is limited not only by class, by minority control of the means of production, but by the division of knowledge. Dunn identifies this as the 'central paradox – that we have all become democrats in theory at just that stage of history at which is has become virtually impossible for us to organise our social life in a democratic fashion any longer' (1979, p. 28).

However, within the technical divisions of labour of institutions, and among groups with similar interests, there is more scope for democratic determination. Even though organizations may be hierarchical and bureaucratic their design and rules may still be subject to democratic control.

Democratic processes are strong on substantive rationality but weak in terms of formal rationality, for they do not oblige voters to compare relative scarcities and budget for what they vote for, and hence to vote for what can be afforded. And given that they are generally voting for public goods, there is a free-rider problem, for voters can rely on the costs of providing a particular asset or service they would like being spread across other voters. (This is not to deny that people should sometimes pay for services used only by others). Governments or institutions in which democratic procedures are used could publish the costings of alternatives so that voters would be able to weigh the costs as well as the benefits of what they were being asked to vote on, but they do not necessarily have an interest in providing this information. They may fear that this would always favour the cheaper option even though sometimes the more costly option may be worth paying extra taxes for. It may help parties get elected if they pretend that they can deliver without sacrifice.[17] Further, there are always alternative ways of paying: should the rich pay more? should the defence budget be cut? should the arts budget be sacrificed? There are invariably costs to whatever one votes for and democratic processes can only provide a crude assessment of them across a limited range of alternatives. Voters might want policy x only on condition that a particular group pays for it through taxes, or that a particular kind of existing expenditure is cut, and they might prefer to forgo it if it is not financed in this way. On the other hand, democratic procedures at least allow decisions on how things should be financed to be made as matters of principle, even if they are weak at ensuring allocational efficiency. In markets, it is not principle but profit and price that determine development; as a consumer you can choose to buy good x and this influences the overall distribution of resources, but the influence works blindly as an unintended and unknowable consequence of your purchase. You don't have the power to decide how resources in general are allocated, only how to spend your own money. For liberals, this is as it should be, for you – and not others casting their votes – are the best judge of your interests, and similarly you and no one else should be responsible for your own property. Clearly, this assumes an atomistic view of interests and fails to acknowledge that many goods cannot be provided through markets because of conflicts between individual and collective interests.

People are not automatically empowered most by democracy; for

---

[17] As Dunleavy argues, although these public-choice arguments have been associated with the political right, they don't have to be (Dunleavy, 1991).

example, being able to sit on committees and take decisions on how resources should be allocated by the state suffers from the problems that the representatives do not know their constituents' needs as well as their constituents do and may have difficulty representing their diversity. (Direct, participant democracy is extremely costly and suffers from the problems already noted.) If people cannot get what they need from markets because they lack money, the solution may not always be to provide what they need outside the market; it may be better to give them the money to be able to use the market and spend according to their individual needs, choosing what they want from competing suppliers. (Note that the suppliers don't have to be capitalist firms.)

By definition, democratic decision-making is decentralized, but the policies and actions that result generally require centralized power to implement them, and this may become a weakly accountable powerbase in its own right. The policies may also suppress local difference in so far as majority decisions are binding on the whole electorate. It is perfectly possible for democratic procedures to produce thoroughly illiberal policies which persecute minorities. To avoid the 'tyranny-of-the majority' effect, a case may therefore be made for setting limits to the scope of democratic procedures (Mill, 1859). In this context markets have some advantages over democratic control. Minorities may have enough purchasing power to elicit a supply that meets their special needs. Even if, under democratic procedures, they were allowed their special provision by the majority, this might be argued to be inferior to gaining it through the market, because the former course implies paternalism and a loss of autonomy.

These kinds of criticism might seem objectionable to democrats, indeed even to note the limitations of democracy might seem shocking and dangerously reactionary. But democratic social choice processes are fallible like any other. Just because a policy is intrinsically good, it does not follow that one can forget about whether it can be afforded. The 'feel-good factor' associated with 'democracy' should not be allowed to obsure its defects.

These three modes are the most commonly noted alternatives to market coordination. Both markets and these alternatives rely upon wider ordering processes and relationships – mutual understanding, formal and informal or tacit rules and conventions, and the successive responses of actors to each other's actions. These are a condition of the possibility of all social life, and their coordinating effects extend beyond production and consumption. Much of social life goes on outside these modes of economic coordination. Markets may have impressive powers of coordination with respect to the production and distribution of goods, but aside from their influence on who gets what, markets only influence the use of those goods indirectly. When we are using our commodities we are doing so outside markets. Even in the most developed market economy, with the most extensive commoditization, goods, including commodities, are for most of the time bought for their use value, rather than their resale value. This applies not only to individuals' property but that of institutions and enterprises. The influence of markets may be pervasive, but most of social life in market economies is not coordinated by the market, either because it doesn't need coordinating or because it is coordinated and regulated by other means. The same goes for planning, democratic control and formal networks, as for markets. Much of the order of market societies derives from similar sources as the order of non-market societies – from norms, rules, conventions and constraints. The romance of the market gains some of its plausibility by mistakenly taking credit for order produced by other means.

Conclusions

In discussing some of the main features of markets, and developing a critical analysis of various concepts thereof, I have moved a long way from the familiar simplistic accounts associated with mainstream economics. The simple terminology of markets conceals many different referents and ways of seeing, and rarely noticed slides between them, often with hidden ideological significance. In many everyday practical contexts, these may not matter, though given the ideology of the 'romance of the market', it would be better if they were understood. The conceptual confusion is certainly a major problem for a political economy which aims to attribute causal responsibility (and hence blame or credit) for outcomes.

I have also more briefly outlined key features of three other important modes, which can both support and substitute for market coordination, though with different effects. In highlighting both their general characteristics and their differing selective strengths and weaknesses, the specificities of markets are thrown into stronger relief. At the same time, all the modes confront problems of coordinating the divisions of labour and knowledge. As we saw in chapter 3, while the divisions themselves can be modified by the modes of coordination, they retain some autonomy and hence cannot be reduced wholly to products of those modes. However, the task of demystifying the markets is far from complete. There are still some celebrated theoretical issues and debates about their merits or evils, and it is to these I now turn.

5

Markets: Key Theoretical Debates and Evaluations

So far I have set out some of the key characteristics of markets and other modes of coordination, modifying both liberal and radical views in various ways. But debates about markets are also conducted at a deeper level in terms of the invisible-hand thesis, ideas of consumer sovereignty, and the efficiency properties and ethical implications of markets. Much is at stake here. Arguments both for and against liberalism and libertarianism conventionally give great prominence to these issues – so much so that for the New Right 'the market' provides the model of social organization. Various claims are made to the effect that the price mechanism coordinates self-interested economic actions more effectively than attempts to coordinate economies deliberately, that this is the most efficient way of allocating resources, that market systems allow consumers to dictate what is produced, that the rewards and resulting patterns of inequality are fair or at least not unjust. Radical political economists generally want to challenge such claims, though supporters of market socialism give some of them qualified acceptance.

In this chapter, I will discuss five fundamental issues: (1) the invisible-hand thesis and the motivational qualities of markets; (2) claims for consumer sovereignty; (3) markets and equality, neutrality and morality; (4) allocational efficiency and markets; and (5) the Panglossian assumption. All I aim to do is discuss the implications of certain features of markets, particularly those which make them different from other institutions relevant to economic behaviour. While I will make various evaluative judgements about particular features of markets, there is no attempt to provide a comprehensive review so as to arrive at a general verdict on whether markets are a good thing. The latter would be of little use (a) because the strengths and weaknesses of markets are likely to vary according to the particular goods or services being traded, and contextual features such as forms of regulation; and (b) because, as I argued in chapter 2, we would have to compare markets with alternatives to arrive at useful judgements.

Before beginning, two methodological points are in order. The first concerns the relationship between positive and normative discourse. Discussions of how markets work inevitably overlap into *evaluations* of this form of social organization because positive statements (about what *is*) in economic matters include allusions to human needs, and hence what is good or bad for us. The relationship between positive and normative here is not one of logical entailment, but the former is at least relevant to discussions of the latter and suggests or secretes certain value positions via implicit or explicit views of human needs (Taylor, 1967).

The second methodological point concerns the fact that this is an *abstract* analysis of markets, not a concrete analysis. In actual markets, there are many diverse influences upon actors, some of them coming from non-market spheres of action in which they are also involved. Philosophers and economists tend to think of markets in abstract terms, sociologists and anthropologists more concretely, especially in terms of the social 'embedding' of economic actions. I don't think these are incompatible, but they are easily confused, with the result that the two sets of disciplines frequently argue at cross-purposes. An abstract proposition about one aspect of markets should not be treated as a prediction of behaviour in concrete markets, nor should it be rejected on the grounds that concrete behaviour is different, any more than we would reject the law of gravity on seeing an aeroplane fly. To challenge the abstract proposition we would have to show that it was not even an element of the situation. Implicit in this argument is an important point about causality. If I push a door and it won't open because someone has locked it, this doesn't mean to say that I wasn't pushing. In concrete situations, non-market mechanisms may compensate for or override market mechanisms, but that does not mean the market mechanisms were inactive.[1] We therefore expect differences between the behaviours which markets *qua* markets encourage, and the actual behaviour of concrete actors. Thus the pressures of competition in a market might encourage an enterprise to drive a hard bargain with a supplier, but this incentive may be overriden or ignored in a concrete market because other pressures are stronger; a supplier who is a family member may be given special privileges. While a study of a concrete case like this would examine these complications and look at the interaction between market pressures and kinship obligations, I am primarily interested in the abstract analysis of what markets as such encourage.

[1] This conception of causality comes from realist philosophy – see Sayer (1992a).

Recalling our discussion in chapter 2 of behavioural assumptions in political economy, it should also be noted that an abstract analysis of markets does not have to assume that all economic behaviour is based on rational choice. Actors may not consciously follow the incentives and pressures of the market, indeed rational evaluation of costs and benefits tends to be limited: for much of the time people act on the basis of habit (though this may not be economically irrational); imitate others, and follow norms or moral principles. But markets do not merely present actors with options which they can then take or leave without further consequence. Market mechanisms also respond to their behaviour, however it may be motivated, and reward or penalize them. It doesn't matter whether your decision to spend all your pay in one go was carefully thought out in terms of costs and benefits or done unintentionally: your lack of money still limits what you can do next. Those whose behaviour is *consistent* with market incentives tend to be rewarded – even if it isn't motivated by them – while those whose behaviour is inconsistent with them are penalized, possibly by bankruptcy or hardship. For this reason, while actors' intentions are relevant to many of the following issues, they are not necessarily decisive. In fact this very point is crucial to the first of our issues.

The Invisible Hand and Motivation in Markets

Central to the invisible-hand thesis is the question of motivation in competitive markets. Producers – whether public, private or cooperative – who are threatened by competition are said to have more reason for trying to satisfy consumers than producers who have a monopoly. A moderate version of the invisible-hand thesis, would claim that through the pursuit of their self-interest in maximizing returns, competing producers are often led to improve the value for money they offer over time (as in the case of computers); producing what customers prefer yields higher returns than producing what they don't want to buy. Without any *necessary* concern for consumers' welfare, but just by looking after their own interests, producers are led, as if by an invisible hand, to produce what their customers want. This is not to deny a role for user 'voice' in influencing producers, but as we have already seen it may be more effective if backed up by the threat of exit as this is likely to impinge most painfully on the producer's self-interest (Hirschman, 1970). Liberals naturally favour competitive-market provision (i.e. financed by revenue from consumers) because this makes a direct link between suppliers' incomes and the extent to which suppliers satisfy consumers. On this view, if economic activities are financed by budgets from taxation, then this link is absent, and especially where budgets are soft, producer interests in empire building and laxity will be indulged by the state and inefficiency will result.[2]

First, let us eliminate one possible objection based on a misunderstanding. We have already pointed out that markets need considerable amounts of labour to make them work and are socially embedded and regulated. Does this mean that the invisible hand is stayed? The answer is no in most cases. The labour involved in running markets can influence buyers' and sellers' behaviour but it rarely determines it or controls what is bought from each seller. Sellers may think about customer interests but prices and flows of exchange value determine whether they survive. It remains the case that prices and hence profits are likely to be highest for goods in short supply and lowest for goods that are unwanted. Regardless of whether suppliers know or care about consumer welfare, their own financial interests will be met best by producing more of the former and less of the latter, thus benefiting consumers. The large numbers of people involved in selling products does not alter this logic. As we noted earlier, even IBM, with its vast numbers of sales and marketing staff, could not overcome the invisible hand in the eighties when technologies and demand shifted (Morgan and Sayer, 1988).

Similarly, political regulation and mediation do not necessarily eliminate the invisible hand; for example, however politicized the competitive struggle between Japanese, European and US producers over markets for consumer electronics and cars, the results still depended on consumer responses to shifts in cost and quality advantages among the competitors (Morgan and Sayer, 1988; Cawson et al., 1990). If the Japanese had not threatened through price and quality, there would have been no need for European and North American producers to organize against them, and hence no need for Japanese to organize in response. Even in this case, the invisible hand was not eliminated, only constrained.

There are several further responses to the standard invisible-hand thesis. One concerns its application to capitalism, as opposed to non-capitalist market societies. The argument that dependence on consumer spending motivates suppliers neglects to note what the internal structure of the suppliers is like, and hence overlooks the fact that under capitalism it is the owners and most powerful members of the firm rather than the ordinary employees who are likely to pocket the gains. While the employees' incomes are ultimately dependent on consumers buying what they produce, the relationship is indirect, being mediated by the capitalist. If workers were allowed to sell directly to customers so that their

---

[2] Neo-liberalism has prompted many experiments in creating quasi-markets in budget-financed public-sector activities by making institutions compete for their funding.

incomes depended directly on consumer spending, the redundancy of the capitalist would be exposed. Capitalist social relations of production depend not only on minority ownership of the means of production but on the ownership of the product.[3] The capitalist does not hire out plant to workers so they can produce commodities directly themselves. To do so would not only allow workers to work under their own direction, but allow them ownership of their products and hence access to their full value. One of the ironies of the invocation of the invisible-hand thesis in the defence of capitalism, rather than petty commodity production or market socialism, is that under capitalism the mechanism is distorted by the mediation of commodity production by capital. To be sure, capitalists' profit is highly sensitive to consumer spending, but consumer purchases are only a necessary condition for capitalist profit, not a sufficient condition. For the capitalist *qua* capitalist, only ownership of the workers' product, which they are able to claim through the power that derives from the ownership of the means of production, enables them to gain access to sales revenue. Workers could turn the tables on bosses who extol the virtues of the invisible hand by offering to improve it by claiming what they make as their own and selling it directly to consumers, so that they feel the full force of shifts in consumer spending, removing the unnecessary mediation of the capitalist *qua* capitalist. Even sales-related bonuses for workers would not remove this problem entirely, as long as mere ownership of the means of production and profit allowed non-working capitalists access to sales revenue.[4]

Aside from noting this 'oversight', a further standard radical response is that motivation is neither so narrowly self-interested nor so limited to financial rewards as the thesis implies; workers can also gain satisfaction from doing their work well and by winning the respect of fellow workers.[5] This is almost certainly so, but it does not confront the liberal argument that such motivations may be at the expense of consumer welfare unless consumers have the option of exit as well as voice. Adam Smith's point was that even though producers who get their money irrespective of the service they provide to consumers may not be lazy and may actually work quite hard, they are likely to work in ways which satisfy themselves rather than the consumer.[6] Radicals may reply that this still fails to acknowledge the power of altruism, for workers are

---

[3] See chapter 6 for further discussion of this point.

[4] Obviously, where capitalists do contribute labour to the production process, the applicability of this criticism is reduced.

[5] Burawoy's work, for example, shows a more complex picture of satisfaction and esteem gained from playing the system (1979).

[6] See Smith's devastating comments on university lecturers of his day! (Smith, 1976, vol. 2, p. 284).

capable of being concerned with the welfare of their customers or clients, or perhaps would be if given more control over their work. This is of course arguable; it is more credible for work that involves elements of consultation, caring and counselling through face-to-face contact between producers and users than distant, anonymous transactions with vast numbers of customers. The consequentialist justification of the invisible hand is compelling when applied to advanced economies where there are too many consumers or users in different situations for producers to be able to know how to assist them without feedback from consumer spending across competing options. Liberals can still with justification argue that even if altruistic motivation is possible, it is inferior in complex market societies to competitive self-interest in delivering the goods. This is the *point* of the invisible-hand thesis (Smith, 1976, vol. 2, pp. 18 and 475).

Altruism may also be unwelcome since it has similar defects to paternalism; the donor is in control and able to dictate terms, and consequently the recipients may prefer to be in a position to choose and pay for themselves. By definition, altruism requires no reciprocal act from the recipient, but donors need resources from somewhere, and their altruism is likely to create a sense of obligation in the recipient. While markets actively corrode altruism, they also avoid creating unwanted obligations, and may actually produce superior goods and services. However, there is more than a hint of circularity in this criticism of altruism, for negative views of altruism and mutual obligations could be partly a result of socialization in highly marketized societies. Nevertheless, noting this circularity does not serve to clinch the argument against markets and the invisible-hand thesis, for it does nothing to advance a case for the *feasibility* of micro-economic coordination of advanced economies on the basis of altruism rather than competitive markets. And even if one wanted to argue for a more primitive economy in which divisions of labour would be simple enough for altruism to be more effective, the associated systems of mutual obligations might be suffocating rather than liberating. Liberals prefer the mobility and individual liberty associated with market societies to what may sound appealing under the name of altruism but which might actually be a communitarian gridlock of mutual obligations. There are interdependences in market societies too, of course, but to a certain degree individuals can change and choose the particular dependences into which they enter: this is the prime attraction of markets for liberals. Just how much choice they have depends on the kind of market and on the number and variety of alternative suppliers; in the case of labour markets, the majority of workers are not in a position to choose to work for themselves instead of as wage labourers.

To draw attention to the limitations of appeals to altruism as a basis for coordination is not to endorse the liberal rhetoric of the ability of markets to encourage 'self-reliance', for this can easily disguise selfishness, misanthropy and contempt for the weak and the unfortunate (see Keat and Abercrombie, 1991; and Heelas and Morris, 1992). It is perhaps these associations that radicals most dislike in markets. Markets may liberate individuals from certain social ties (though not, as we have seen, from social relations) and from the burdens of democratic involvement, but they can foster a kind of amoral individualism which disregards the social consequences of both the circumstances under which they are able to acquire products and services and the effects of their own acquisitions. Commodified education and health services allow those with sufficient money to buy the services without any justification to others who have equal need for them but from whom resources may be diverted. They allow individuals to bypass the whole question of whether certain things ought to be provided regardless of ability to pay. Economic liberals celebrate markets because they appear to embody the realization of the principle that as far as possible individuals should be free to pursue their own conceptions of the good. As some critics have argued, in cases like those of health and education, markets do not realize this ideal and actually seriously restrict some individuals' freedom (Haworth, 1994). There may be dangers of paternalism in state provision, but right-wing claims that markets encourage self-reliance instead of dependence tend either to ignore inequalities in the resources needed to be self-reliant or overlook the question of whether those inequalities are deserved and justifiable or a consequence of exploitation, beggarthy-neighbour behaviour, or luck – all of which are common in market situations.

Consumer Sovereignty

Closely related to the invisible-hand thesis and central to liberal celebrations of 'the market' is an emphasis on consumer choice. The Left has traditionally been suspicious of notions of 'consumer sovereignty', but while its criticisms have considerable substance, they are often exaggerated, sometimes to the point where consumers are represented effectively as dupes. Throughout the following discussion it should be remembered that consumers include industrial and institutional consumers as well as the individuals generating 'final demand':

1   The whole emphasis on the liberty of the consumer serves to distract our attention from the sphere of production where workers' liberties and control

over their work are heavily curtailed. Consumer welfare is not all that matters. Consumer demands may be met by exploiting child labour, by exposing workers to dangerous chemicals, or by creating unemployment.

2   Celebrants of consumer sovereignty often fail to acknowledge that consumers are highly unequal, both among themselves and with respect to sellers. It is especially misleading to compare markets with elections as if everyone had the same number of votes with which to choose products. The poverty of many consumers fits ill with the connotations of 'sovereignty'.

3   Another reason for opposition to the notion of consumer sovereignty from the Left and from sociology is its association with an undersocialized view of individuals as free, autonomous actors (Warde, 1994). We can accept that consumers are not asocial individuals, but are embedded in relationships and are subject to cultural influences, so that we can therefore roughly predict what certain kinds of people will buy according to their 'habitus' (Bourdieu, 1984). Markets are not the servants of autonomous 'preferences' but influence them, albeit often in unpredictable ways. But the presence of these influences and expensive advertising campaigns does not mean consumers have no choice. I may have been socialized into eating muesli but I can choose to buy something else. People – including academics! – are sensitive to the influences of culture and advertising, but they are not dupes.

4   Some radicals may be tempted to be simply dismissive of the fact that one can choose between a vast array of commodities, but though it may not impress them, the net effect of their choices has major implications for the lives of those who produce the commodities. At least in competitive market systems, consumer spending had a direct effect on producers, unlike in state socialism where it need not. The lack of consumer power under state socialism contributed much to the shift in opinion in favour of 'the market'. This suggests that it is foolish to belittle the significance of consumer choice in competitive markets.

5   Whether consumers are sovereign depends, among other other things, on their access to information – information which it may not be in suppliers' interests to provide. Real markets, embedded, regulated and characterized by imperfect and unevenly distributed information, are effectively actual or potential fields of struggle. Hence, as Elson (1988) emphasizes, markets don't have to be simply accepted on existing terms or rejected; they can be further socialized and politicized, and behaviour within them changed.[7]

6   Often the power of the consumer is compromised by 'natural' producer monopolies of information. In some cases, for example medicine, consumers do not or cannot easily know what is best for them. Alternatively, in the case of information products, ignorance of what they contain is a necessary condition of wanting them. Students cannot get much idea of the content of a course until they have done it, and they would not need to do it if they already understood its content.[8]

---

[7]   My thanks also to John Lovering for impressing this on me.

[8]   Education implies a kind of inequality between 'supplier' and 'user' that makes market criteria inappropriate. Nevertheless, this does not exempt education from assessments of

Before moving on to the implications of markets for equity, it should be noted that while we have made numerous, quite major qualifications to idealized accounts of markets, we have not tried to deny – as do some radicals – that markets are competitive or that consumers have some degree of choice, at least compared to other modes of coordination. Rather, we have argued – both here and in the preceding chapter – that competition and choice work in different ways to that assumed in standard accounts.

Equality, Neutrality and Amorality

Controversies about markets invariably get into debates about their effects on equality. Competitive markets have both egalitarian and inegalitarian tendencies. As Hayek admits, success in markets may have nothing to do with merit or fairness and much to do with luck; markets are neither meritocratic nor egalitarian institutions, though they *sometimes* reward merit and have egalitarian effects. Ethical issues naturally come to the fore here, and while I shall make some comments on these I am primarily concerned with the ways in which inequalities are produced. There is not space here to go into the question of whether or how much equality should be valued, but it should be noted that I do not wish to assume any blanket condemnation of inequality.

The simplest inegalitarian market mechanism works through consumer spending; if most consumers opt for A's product rather than B's, whether on price or quality grounds, then inequalities will open up between the incomes A and B receive as an unintended consequence of consumer choice. Although this may prompt B to change to a more remunerative occupation, unless and until that happens inequalities in income will persist. This is how the price mechanism and the invisible hand are *supposed* to work; without the inequalities there would be no incentive to change to producing goods which consumers would prefer. Note also, that contrary to a common impression in radical thinking,

---

opportunity cost. Other common radical criticisms include the claim that there is a ridiculous excess of different versions of commodities under capitalism (do we really need so many kinds of jeans or gadgets?), and that competition is wasteful, or else that the choice is between brands that are all too similar, hatchback cars being a favourite example. Neither of these arguments makes much headway against the point that competitive markets give consumers choice. The first argument implies that there is too much choice and the second could be construed as implying that there is not enough, though it fails to say why, if consumers wanted different cars, it would not be profitable for producers to make them. Moreover, lack of variety could reflect the fact that consumers may not want to be different.

this mechanism operates through the unintended consequences of market choices and not through relations of domination.

Further inegalitarian mechanisms work on the production side through ownership of means of production, and these involve exploitation in the sense of some individuals living off the labour of others. Here, the work of the analytical Marxist John Roemer has been especially useful. One such mechanism works through market exchange independently of class relations. Imagine there is a number of independent commodity-producing enterprises, each with its own means of production. Unless the enterprises have equal assets in means of production, some producers will expend more in labour time in producing what they sell than they get back in terms of labour embodied in the goods they buy, and those with plentiful or highly efficient means of production will be able partly to live off the labour of others. Exploitation will therefore occur. This mechanism can obviously work where the enterprises are capitalist, but it operates not through capital–labour relations but through exchange relations. It is not dependent on capitalist class relations, or indeed on any relations of domination, but is a consequence of inequalities in holdings of means of production (Roemer, 1988). It is therefore likely to occur in non-class market systems such as market socialism (as we have defined it) or in a petty commodity production system where the enterprises are self-employed individuals.[9]

Another mechanism also arises from inequalities in holdings of means of production, but this time involves the development of wage labour and hence class. Imagine that all enterprises are self-employed petty-commodity producers. Those producers whose means of production are either too meagre or too obsolete to allow them to earn a living in competition with larger, more advanced producers may be *better off* working for those with large holdings, and hence they may *opt* to become wage labourers (Roemer, 1988). They will nevertheless be exploited in the sense that they will produce goods whose value exceeds that of the goods they can buy with their wages, the difference going to their employer. It is also possible that an initial inequality in holdings of means of production could come about through different preferences for saving and consuming. If some petty producers prefer to hold down their consumption more than others, and invest their savings in additional means of production, they may be able to employ others by offering them more than they could earn on their own (Elster, 1985, p. 226).

These appear to be 'clean' or voluntary routes to exploitation, involving an element of choice, but they are unlikely to remain so. Once

---

[9] Inequalities and exploitation can also occur between enterprises as a consequence of differences in their capital–labour ratios.

the employment relation is established it allows further exploitation and capital accumulation via the mechanism emphasized by Marx. Further concentration of means of production into the hands of a minority makes wage labour *unavoidable* for the majority rather than a matter of preference. Owners can then live off the labour of others without further necessary personal sacrifice, and their children can inherit their gains without lifting a finger. While it is possible for exploitation to originate cleanly, this does not mean it remains fair; by the same token, it does not follow automatically that because exploitation in the Marxist sense seems unfair, that it can never have clean origins.

These situations are interesting in that they show how market systems which initially need not involve class are liable to convert into class systems involving exploitation within production itself alongside exploitation through unequal exchange. Expropriation is not the only route to proletarianization. To avoid these spontaneous conversion processes, it would be necessary to ban capitalist ownership and insist on state or cooperative ownership of all but the smallest enterprises.

There are other mechanisms affecting equality that are associated with markets but which are more complex. To appreciate these we need first to note the *neutrality* or *indifference* of market processes. Other things being equal, markets tend to work 'without regard for persons', as Weber said in another context. In the logic of markets, all that matters to actors is what they have to sell, what they want and can afford to buy, and costs and prices. If we respond only to these things, then we will be *neutral* with respect to other characteristics of the actors. Behaviour in concrete markets may be influenced by other considerations, such as whether actors happen to like the person they are trading with, but these are not essential *market* considertions: many things can happen in concrete markets, but not all those things are products *of* markets. Market incentives and pressures are nevertheless powerful, indeed one of the main objections to markets is that they tend to override the non-market considerations so that social interaction is dominated by the essential requirements of the cash nexus: life itself in a highly developed market economy is 'economistic'.[10]

---

[10] The word 'essential' will set alarms ringing in anti-essentialist circles, but clearly we are making an *abstract* claim here. In concrete situations, which are outcomes of interactions between several processes, other considerations are likely to be present too. If we can't use abstractions, then it is hard to imagine how we could think and act at all, for we could never distinguish between things which, though often combined, are separable (and hence can be considered through separate abstractions), and things that are not. And if things have no essences, such that there is nothing to abstract, then we have to assume anything can happen to anything, and thus that there is nothing that we can expect to achieve through our actions.

This neutrality derives partly from ignorance: we do not know (and often do not need to know) much about the circumstances in which the products we use were produced. Marx's concept of commodity fetishism denotes the way in which the producers of commodities and the social relations under which they work are invisible to the purchaser, whose economic relations appear to be with things – the commodities themselves – rather than people. However, ignorance of the social characteristics and circumstances of the makers and users of products is not unique to markets. It is partly a function of the fact that products rarely bear the mark of their producers' character, and that where the division of labour is fairly developed, producers and consumers are likely to be distant from and unknown to one another, regardless of whether they are coordinated by a market or some other means. Someone who gets medicine allocated to them free of charge from the state is likely to know no more about their social relations with its producers than someone who buys it in a market. Under an advanced division of labour, some of the effects referred to by Marxists under 'commodity fetishism' can therefore be found where it is use values, rather than commodities, that are being distributed. Money and markets do have additional effects, though; to the separation and mutual ignorance produced by division of labour, the use of money (even without competitive markets, as in state socialism) adds another kind of indifference: a pound is a pound whoever it comes from, wherever it has been. Yet capitalism is not the only economic system which has commodities and hence commodity fetishism; a pot of paint bought from a state store under communism and one bought in a market may be identical and each will tell us equally little about its producers.

Ignorance about the social origins of products is not always a problem. Market coordination of the economy saves us a great deal of time because of its low information requirements. However, opacity is a problem if it conceals social or ecological exploitation, for it prevents us adapting our economic behaviour to take account of non-economic considerations to which we would like to respond. I cannot tell from the taste of an apple whether it was grown by a racist or a non-racist. I may wish to know who grew them so I can avoid helping those I wish to disfavour, but I cannot usually deduce this from the nature of the product alone or its price. Likewise we often do not know where those with whom we trade obtained their money. Even if we do, and we decide that somebody's money is 'tainted' – for example, the ill-gotten gains of the arms dealer – the money itself does not actually lose any value through its associations with particular holders, and so others may be tempted to accept it. Competitive markets add a further incentive to be neutral and to refuse to make a moral stand on particular products, producers or consumers, for fear of losing ground to competitors who are neutral. Thus enterprises in a competitive market are tempted to break sanctions in order to get an advantage over others, or at least to avoid losing to others who ignore them. In other words, market economies – not just capitalist ones – combine *several* sources of neutrality or indifference.

Earlier, in our discussion of the relationships between capitalism, race and gender in chapter 3, we touched on the relationship between the neutrality and egalitarian and inegalitarian qualities of markets. These kinds of qualities can be expected in non-capitalist market economies as well as in capitalism. For example, in labour markets, there are both egalitarian and inegalitarian mechanisms; markets can discourage discrimination against particular kinds of worker, since those who discriminate on grounds irrelevant to their ability to do the job have a smaller pool of labour from which to choose and therefore lose out to competitors who do not. On the other hand, concrete markets operate in environments in which there are invariably already inequalities and discrimination arising from other, non-market sources. There are consequently some cases where actors in market situations may benefit by reinforcing certain inequalities and types of discrimination that exist in society. Some suppliers may find an economic advantage in racist employment policies in response to their customers' racism, so that predominantly white clients are not served by blacks. In housing markets, suppliers often segregate housing in order to protect property values, thereby reinforcing racism. If people are unequal on arrival in the market-place, then the inequalities will be reproduced or increased through market transactions. In chapter 3 I argued that despite the fact that capitalism had adapted to and taken advantage of patriarchy, racism, homophobia, exploiting the resulting inequalities where possible, albeit not without major difficulties, there was no reason why it could not survive without them. Even though certain enterprises might only be able to survive because of the existence of a disadvantaged group which they can use as a source of cheap and malleable labour, capital in general does not depend on such contingencies, because there are many other means by which firms can compete and survive.

Equal-opportunities policies in employment generally follow the liberal logic of market neutrality. They counter discrimination on grounds other than ability to do the job – applicants for an accountancy job should be judged on their ability to do the job, not on irrelevant considerations, such as their sexual orientation. Yet such a policy does not question the nature of the job itself and its suitability for people who, though they may be equal in ability and training, have different constraints. Thus a professional job requiring over 60 hours work a week is effectively a job for those who don't have young children or heavy domestic responsibilities. Employers advertising such a job could claim that they only discriminate in relation to ability to do the job, but they may just as well put on the advert 'single parents and others with heavy domestic responsibilities need not apply'. Individuals entering markets are unequal in their constraints. It is not only the selection of people in the labour market which can lack neutrality, but the jobs on offer. Equal-opportunities policies and market logic might counter discrimination in access to opportunities but they do little to combat bias in the opportunities themselves. They fail to challenge the liberal model of the individual – implicitly an adult man unencumbered by responsibilities to care for others – and the way in which this biases the construction of opportunities. Similarly, there are biases in the housing and holiday markets towards heterosexual couples and families and against single people, in that while access to the products may be open to all, the products tend not to be designed for singles or same-sex couples. But in all these examples, although the bias in the products on sale is reproduced through markets, the responsibility for the bias lies largely in extra-market forms of discrimination – patriarchy influencing the design of jobs, cultural discrimination against singles and homophobia in the other example, etc. What appears to be a bias in these markets is arguably an effect of their neutrality in reflecting other sources of inequality and discrimination. This conclusion is reinforced by the probability that in a competitive market, as long as the excluded groups had sufficient purchasing power, there would be an incentive to someone to step in and sell them the goods and services that they wanted.

It is important to realize that these points about neutrality and discrimination apply also to non-capitalist market systems, such as market socialism; they are not dependent on the capital–labour relation. Hence, although we might hope that these would not be racist, sexist or homophobic, the same combinations of egalitarian and inegalitarian tendencies or potentials would be present as under capitalism, by virtue of their dependence on competitive markets. Moreover, as with capitalism, while racism, sexism, etc., might in fact be common in such workplaces, they would emphatically be contingent rather than necessary features of cooperatives.

The contingent nature of these relationships between markets and these various forms of discrimination makes it possible for proponents of market socialism and Left-inclined liberals to respond to them by arguing that it's not the market that is at fault here but non-market forms of discrimination: eliminate the latter, and the only things which will count will be ability to pay and the quality and price of what you have to sell, be it a product or your own labour power. Such a line might seem alarmingly complicit with a common rationale given by market actors engaged in these discriminatory practices; for example, shop-keepers who refuse to hire ethnic minority staff so as not to deter racist customers may say they are not racist themselves but are merely doing what is good for business.[11] However, the complicity can be avoided by recognizing the difference between what concrete actors should do in market situations and what the market itself encourages them to do. Bearing in mind this distinction it is not unreasonable to argue that market incentives are not inherently racist and should not be blamed for these ills, but that the actors operating in markets *are* culpable and ought to resist racism, even if it means forgoing economic gains. As we shall see shortly, market incentives can't be expected to have any moral warrant, but we do expect the actions of concrete actors to have such a warrant, for as whole people involved in many spheres they have to think of other considerations beyond those of market success.

Sometimes, firms or interests representing firms may be encouraged by market incentives to engage in discrimination *outside* the market in order to give them advantages within the market. They might foster racism to create a stigmatized group which will accept very low wages. One doesn't have to be a conspiracy theorist to note that where substantial economic interests are at stake, practices outside markets can be influenced by their effects on those with interests within markets. Although it is useful to think about markets in abstraction from other institutions with which they may be combined, it has to be remembered also that actors operate in concrete markets and move continually between market and non-market spheres, and may try to orient their behaviour in one sphere towards their interests in the other.[12]

A paradoxical feature of markets is that their neutrality or indifference with respect to individual characteristics is accompanied by an extra-ordinary responsiveness to differences in consumer preferences. (Note again this is still an *abstract* discussion of particular tendencies deriving from the structure of competitive markets; it does not rule out the possibility that they may be overriden by other non-market tendencies in concrete examples). Many products have associations with particular kinds of consumer identities, and sellers will generally benefit by

---

[11] Positive discrimination policies within markets (e.g. lower prices for pensioners, recruitment quotas for ethnic minorities) would introduce a non-market or anti-market logic, in that customers, suppliers and employers and employees would be assessed on grounds other than price and quality. Individual employers are of course free to use positive discrimination even where there is no legislation, but unless enforced universally they could damage their firms' competitiveness.

[12] For an excellent analysis of this spill-over of rationales and interests from one sphere to another, see Walzer (1985).

targeting those consumers and designing products specifically for them. I don't mean the obvious examples, like lipstick for women and shaving sticks for men, but rather the subtle differences between variants of the same product; small, chic hatchback cars are targeted at young women with money, high-powered 'hot-hatches' at young men. On the face of it, nothing could be further from selling 'without regard for persons'. Yet even in these cases, the market logic of indifference still has a place, for any firm which refused to sell a product to someone they had not targeted but who wanted it nevertheless – a hot-hatch to a middle-aged vicar, for example – would risk losing sales to competitors.[13] Within a *non*-market situation, identity frequently comes before ability to pay, whereas in market situations, ability to pay still rules over identity. Entry into a particular subculture can't generally be bought but depends on having a compatible and acceptable identity. As a middle-aged white man, I couldn't buy acceptance as a member of an Afro-Caribbean youth subculture centred on reggae music, but I could easily buy the same music tapes as the members. In these respects then, markets appear to encourage neutrality and allow actors more freedom than they have within non-market institutions; just as money has the power to infiltrate an enormous variety of groups and transgress social boundaries with complete impunity and indifference, so the holder of money gains in liberty and can enjoy in commoditized form things which she might otherwise be denied.

Early defenders of markets argued that they had a 'civilizing effect', reducing the power of the sovereign, encouraging individuals and nations to be industrious and attentive to one another's needs, and to cultivate a reputation for fair dealing so as not to damage their credit-worthiness and market position. Competition between suppliers may be cut-throat, but as Simmel argued, it encourages them to serve the customer with great assiduity:

> Innumerable times [competition] achieves what usually only love can do: the divination of the innermost wishes of the other, even before he becomes aware of them. Antagonistic tension with his competitor sharpens the businessman's sensitivity to the tendencies of the public, even to the point of clairvoyance, in respect to future changes in the public's tastes,

---

[13] In practice, marketing people generally hedge their bets: the young woman's chic little hatchback is occasionally shown being driven by her boyfriend, and he looks no wimp. Consequently, products with multiple symbolic meanings may sell better than those with only one: part of the success of the mini jeep derives from the fact that it can compliment both macho fantasies and feminine chic. One reason for developing multiple meanings is that many products are shared by individuals with different identities – the car may have to be both his and hers.

fashion, interests . . . Modern competition is described as the fight of all
against all, but at the same time it is the fight *for* all.
(Simmel, Conflict and the Web of Affiliations, *cited in Hirschman, 1982,*
p. 1472)

Hirschman comments more generally on the thesis of the civilizing
influence of markets: 'Commerce is . . . seen as a powerful moralizing
agent which brings many non-material improvements to society even
though a bit of hypocrisy may have to be accepted into the bargain'
(1982, p. 1465). While this is an important point, and the 'moral
behaviour' something which we would presumably rather have than not,
it is encouraged by market incentives for instrumental reasons – not
because it is good in itself, but because it pays off. In another sense then,
markets are amoral. The information, penalities and incentives provided
in markets do not encourage people to act morally, except in so far as it
serves market ends. Humane treatment of workers is morally preferable
to inhumane treatment, but under the logic of market competition,
humane capitalism only wins out if it turns out to be more profitable.
The moral qualities of applicants for jobs are often relevant to their
ability to do the work, but the market reason for choosing an honest
and helpful person rather than a dishonest and rude person to deal with
customers is that they are likely to be better for sales, and only
incidentally that being honest and helpful is good in itself – despite what
firms' mission statements may say to the contrary! (Sometimes, dis-
honesty may help of course!) It is not that *many* acts responding purely
to these conditions are *im*moral, but that their rationale is *a*moral.
Indifference to irrelevant qualities like those of sexual orientation or race
in selecting people for jobs can be defended on moral grounds, just as it
can outside markets in marking exam papers or voting, and actors in
concrete markets may actually be motivated by such considerations, but
it is important to recognize that *market* neutrality is not based on these
grounds. It derives either from self-interested consequentialist reasoning
with respect to competitive advantage (what will be best for my income/
profits?), or through the effect of competition in penalizing those who
fail to recognize or follow such reasoning (those who discriminate lose
in the market). In practice, as we have seen, there are good reasons also
for expecting discriminatory practices within markets where they coexist with discrimination outside markets.

The neutrality of market logic might be a good thing where indifference is morally warranted, but it is not where there are moral grounds
in favour of discrimination. If I knowingly buy goods made with child
labour on the grounds that the product is as good as those made by
adults, or sell drugs to children on the grounds that their money is as

good as anyone else's, then I am acting immorally. Market incentives do not distinguish between immoral and moral earnings. In concrete cases, actors may ignore the market incentives and act on moral grounds, but that does not mean the market incentives do not exist.

Note though, that it is easier for those who are not dependent on the transaction to discriminate than those whose livelihood is dependent on it. I can easily stop buying tropical wood so I do not contribute to deforestation, but it is not so easy for someone whose income depends on logging to stop it. *One of the most fundamental criticisms of markets is that they dominate individuals' lives to the extent of making it unreasonable to expect them to act morally.*[14] Where those whose lives depend on success in competitive markets do claim to be acting on moral grounds, we are apt to respond cynically by concluding that they are only doing so because it is not hurting their income. Thus a cosmetics firm can refuse to sell products tested on animals, but there is always likely to be a suspicion that they are only doing this because they can get more sales by advertising this fact, or at least that it doesn't lose them any.

Despite the fondness of neoclassical economists for the distinction between positive and normative matters, they are apt to allow the latter to colour their judgements of positive questions, so that they tend to explain the existence of markets in terms of what is supposed to be good about them. Markets can operate in situations which we might judge as fair, such as that of the prisoners exchanging gifts, and others which we might judge as unfair, where people trade under duress. The fact that the former kind may be judged to be desirable, does not mean that all markets are like this. Markets don't need to be fair to operate, they can run in conditions where most participants are trading under duress. While participants are 'free' to agree to particular exchanges and prices, they may not be free in any meaningful sense to leave the market and make their living by other means (Haworth, 1994). Even if we accept that users of markets are free to make choices within them, it does not mean that they would freely opt for a market system if alternative forms of economic organization were available, though of course nor does it mean that they would necessarily prefer the alternatives.

A more fundamental moral question concerning markets is how we decide whether certain things *should* be commodified.[15] Slavery, or the sale of things of particular spiritual significance, are examples where

[14] Hardship in non-market economies might also pressure people into acting immorally. I do not claim that markets are the *only* economic phenomenon to encourage such effects.

[15] There is clearly a prior issue of property rights here, for once we allow something to become private property, the decision of whether it can be sold is taken out of our hands.

most people would object to commodification in principle. Others, like love, *cannot* be bought. Many are contested: some people might not object to the sale of the pyramids; and doctors in India involved in procuring human organs from the poor have recently defended their practice against alien Western values. But there are also arguments in favour of extending the exemptions from commodification, as in certain green political theory which rejects attempts to put monetary values on the environment (Sagoff, 1988). To refuse to allow anyone to own something like a sacred site or area of great natural beauty, and hence to prevent anyone commodifying it, is to invoke absolute communal values. Yet the invocation of such values implies either consensus or suppression of dissent; it does not fit at all well with multicultural societies character-ized by value pluralism.[16] Christians' or agnostics' values regarding a mosque in a British city are likely to differ from those of a British Muslim. Even within a single culture, divisions of knowledge and labour themselves undermine communal absolute values. Communitarian values may be repressive and intolerant of difference, so that cultural pluralism may actually be more attainable in a market system, given the neutrality of markets. I will return to these issues in chapter 9.

On balance, the regressive tendencies of markets tend to outweigh the egalitarian tendencies. However, noting this does not amount to a knock-down case against markets, even if we agree that the inequalities are a bad thing. There are other aspects of markets to consider beside their effect on distribution, and some of these might be regarded as compensating virtues. Furthermore, markets are not the only source of regressive economic processes. Where fears of regressive effects of markets encourage a policy of free provision at the point of delivery as a universal benefit, this of course merely dispaces the costs elsewhere, and, as LeGrand has shown, affluent users are often more able to take advantage of such services than lower-income groups for whom they were mainly intended (Le Grand, 1982). Free provision can therefore allow the comfortably-off to be subsidized partly by poorer taxpayers.[17]

The incentives of price and profit work imperfectly and their effects

[16] Nor does it fit with liberal preferences for maximizing individual liberty. Here I am simplifying a very complex field of issues; some liberals do not accept that the claims of individual liberty would trump absolute values in all such cases (see Kymlicka, 1989). Moreover, where practices are debarred from commodification, the issue of whether and how these practices should be supported economically is often contentious (See Williams, 1974 and 1977; MacIntyre, 1985; Keat and Abercrombie, 1991; Heelas and Morris, 1992; Keat, Whiteley and Abercrombie, 1994; Walzer, 1985).

[17] Milton Friedman's slogan 'there is no such thing as a free lunch' is quite correct, but it doesn't deserve to be monopolized by the Right. Leftists and feminists could also use the slogan to good effect against capitalists, landlords and rentiers and against men who live off women's unpaid labour.

may be overriden by those of other mechanisms, but they do have egalitarian as well as inegalitarian effects. As we stressed in chapter 2, the discovery of mechanisms which generate important problems is not sufficient to warrant calls for their abolition, for the alternatives might be worse.

Allocational Efficiency

Many economists believe that the greatest advantage of markets lies in their capacity for encouraging allocational efficiency. As David Steele has argued in his analysis of von Mises' critique of socialist views on economic calculation, the nature of allocational efficiency is rarely understood by socialists (Steele, 1992). The allocation of scarce resources is not simply a matter of meeting ultimate ends, for there are invariably competing ends and different ways of meeting them, involving different expenditures of time and resources. Trade-offs among alternative means or ends can always be made, though often these are not consciously considered. Allocational efficiency is not the same as mere technical efficiency (that is, the minimization of wastage of materials or labour time in producing a given product, or its inverse, the maximization of output from a given set of inputs). Rather it concerns the distribution of resources, including labour power, among different competing ends, and among different means to those ends, taking into account the different intensities of demand or need for each possible product or service, and the relative scarcities of each resource. We have coal, ore, fertile land, labour skills and thousands of other resources, and we have millions of competing possible uses to which they might be put, from public swimming pools to spark plugs to health care to computer monitors to domestic heating, and so on, and alternative ways of producing those things. What would be the best allocation among all the different possibilities?

To simplify, imagine there are two products, A and B, between which we are indifferent as final products. Both are made with high degrees of technical efficiency, but A is made using very scarce resources and B with plentiful resources. In this situation it is easy to see that we should opt for B and save the scarce resources for uses for which there are not more plentiful (i.e. less costly) substitute inputs. Just how the rare resources formerly used as inputs in making A should be deployed instead is not merely a matter of technical efficiency (many applications might score well on this criterion) but allocational efficiency. To assess the 'cost' of something is to compare these allocations and what is forgone by any particular allocation. As soon as these decisions cover more than a small range of alternative resources, products and needs or demands, the combinatorial possibilities become huge and allocational efficiency and cost become difficult to assess directly without the aid of prices as a measure of cost.

Socialists traditionally underestimate these problems. In this respect the precedent was set by Marx himself. For example, in volume 1 of *Capital* (1976, pp. 169–71), he parodies bourgeois economists' portrayal of Robinson Crusoe deciding how to use his labour time by calculating how long different tasks take – though Marx ignores the question of the relative intensities of his various needs. Then, after a discussion of the allocation of labour time in the feudal era, he considers how 'an association of free men, working with the means of production held in common, and expending their many different forms of labour power in full self-awareness as one single social labour force' would allocate their labour power. He concludes that 'All the characteristics of Robinson's labour are repeated here, but with the difference that they are social instead of individual' (p. 171). Engels fared no better, considering it a 'trifling matter to regulate production according to needs', a richly ironic claim in view of the history of central planning (Steele, 1992, p. 27). This not only reduces catallaxy to economy*, as noted by Hayek, conjuring away the immense complexity of a modern economy and its division of labour and knowledge, but also falls foul of the problem of allocational efficiency.

How, without money, prices and markets, can one compare the costs of all the thousands, perhaps millions, of different possible allocations of goods? Market prices, i.e. prices affected by competition, and the flows of income that they stimulate and regulate, are sensitive to these relative scarcities and intensities of demand. (Remember that although firms can in principle fix any price they like, the volume of sales and revenue they get is determined by the decisions of others). Prices and the flows of exchange value shift and respond without it being necessary for anyone to attempt the impossible taks of overcoming the division of knowledge so as to find out the costs of all the possible alternative uses of resources. The Misesian analysis of allocational efficiency is therefore a development of the Smithian invisible-hand thesis, recognizing not only the influence of the decisions of final consumers on producers but the effects of relative costs across the whole input-output matrix of the economy.

Without prices, we would need to know not only the labour times, but the relative scarcity of materials, and the relative intensities of demand for them, and not just the final demand of consumers but the intermediate demands of producers too. It is tempting, following Marx, to reduce all these questions to ones of labour time. But the relative scarcity of two resources is not necessarily equivalent to the amount of labour time expended in extracting or producing them (Steele, 1992, p. 147). The scarcer of two resources does not always require more labour time to produce, and resources – including machines – which take equal times to produce are not necessarily equally productive and hence equally useful.[18] The concept of socially necessary labour time takes account of technical efficiency in terms of labour time required to produce a given unit of a product at a particular level of development, but it does not measure how costly that product or process is in terms of scarcity relative to the intensity of demand for it.

> If the economic problem were merely the engineering one of getting the best use of labour and equipment to maximise a given purpose the problem could be evaded, but once a variety of purposes are admitted this involves the problem of *cost*. This is an economic problem, not an engineering one, because economic cost is understood subjectively as the forgone value produced by alternative uses of a resource.
>
> *(Barry, 1979, p. 180)*

Up to a point, the labour theory of value illuminates exploitation – the way in which some can live off the labour of others. The fact that machinery and technology, as well as labour, can increase productivity and output does not alter this but rather helps show how exploitation occurs; the greater the inequalities in the productivity of the machinery held by different groups, the greater the scope for exploitation. However, exploitation depends also on the scarcity/relative demand for the commodities being produced – no one can live off the labour of others if they produce unwanted goods. The labour time therefore has to be socially necessary in another sense – not only measured in terms of average technical efficiency but in terms of how much the product is needed. For the same reason, the labour theory cannot double up as a theory of economic calculation, i.e. for explaining and prescribing the allocation of resources, whether for capitalism or socialism. Comparing labour times would not tell planners, any more than capitalists, which use of resources to adopt.

This is a serious matter, for if we attain only a low level of allocational efficiency, then waste and sub-optimal rates of development occur, and opportunities have to be forgone. Thus, being able to afford to provide complex medical operations is not merely a question of the technical efficiency of those operations and the associated health services, but depends on allocational efficiency across the whole economy. If labour power and other resources throughout the economy are allocated without regard for both their relative scarcities and the relative intensities

[18] Marx actually recognizes the second of these two points in vol. III of *Capital*.

of demand or need for them, then many otherwise feasible possibilities will have to be forgone, because resources are tied up in applications which are allocationally inefficient even if they are technically efficient (Steele, 1992). For liberal economics, other modes of coordination, such as planning and democratic control, lack the discipline of competitive markets. They are therefore always liable to produce some goods beyond the optimal amount and so cause undersupply of other goods. Thus, bureaucratic organizations are susceptible to pressure for empire-building if they are financed by taxes and are hence not constrained by a mechanism which compares the scarcity and intensity of demand for the goods or services they use up and produce relative to others. Similar arguments are made regarding democratic control and the influence of 'special pleading' by pressure groups (Thompson, 1990).

This, then, is what liberal economists mean by allocational efficiency and why they believe competitive markets produce it. However, even if we accept, as I think we must, that allocational efficiency is an important concern and one that is not reducible to technical or engineering efficiency, there are several objections to the above arguments.

First, in real economies, the allocation is influenced not merely by relative intensities of need or demand but by differences in purchasing power, and by private control over certain resources such as land. We could argue that (some of) these differences are unjustified, and hence that the allocation of resources across market economies is anything but efficient. However, while this highlights an important countervailing tendency, it does not succeed in undermining the analysis of allocational efficiency, for the latter would still matter even without class and income inequalities and with different distributions of non-class power.[19]

Second, there is a macro-economic, Keynesian objection which addresses the absurdity of claiming efficiency for an economic system which routinely fails to employ millions of workers: 'The market mechanism fails to provide a means whereby workers can signal to firms that they would demand more goods and services if only they could get jobs and so have money to spend' (Levacic, 1991, p. 44). In other words, while markets provide signals to which both producers and consumers can respond, there is no mechanism to reconcile the responses to people in their role as producers, where their wages appear as a cost, and their role as consumers, where their wages appear as a source of revenue. It should be noted, however, that the trade-off between keeping costs of production down and keeping consumption up is not unique to market economies such as capitalism, as a 'contradiction'. A planned economy

---

[19] I am omitting further complexities here, concerning justifications of certain inequalities in terms of allocational efficiency. For some of the arguments, see Steele, 1992.

would also have to balance the two. What is specific to market economies is the *form* this trade-off takes, and the inadequacy of market mechanisms for resolving it.

The third point is that allocational efficiency is not the only kind of efficiency, and that efficiency in production and innovation are also important (see Hausner et al., 1992). Here the neglect of production in liberal economies turns out to be a liability. In so far as it considers production at all, it reduces it to allocation, to the simple additive combination of 'factors of production'. But the kind of output that results depends on how their combination is practically organized, on the form and sequencing of operations (Langlois, 1986), on the resolution of space–time coupling problems of various operations, and particularly on the nature of the contribution of labour (including management) – whether it is indolent or industrious, cooperative or resistant. The remarkable development produced by capitalism is far from wholly attributable to capitalists choosing the right inputs and outputs according to market prices. Prices of inputs and outputs do not supply much of the information relevant to the organization or production, the design of labour processes or the selection of management methods, and certainly little guidance in relation to investment and innovation. There are many non-market as well as market sources of information for these matters, including imitation, and the specific know-how needed for producing each type of product also usually requires considerable learning-by-doing.[20] The same argument applies to market-socialist and state enterprises; to the extent that they have been successful this is owed partly to these other kinds of efficiency. Once again, the market optic of liberal economics induces a tendency to give markets the credit for effects produced or co-produced by other mechanisms.

Fourth, allocational efficiency and innovative or dynamic efficiency may also work against one another. Transferring resources – and even enterprises – from less-needed to more-needed activities enhances allocational efficiency, but the associated insecurity hanging over any production enterprise is likely to deter it from making long-term investments involving fundamental innovation. Reducing the mobility of capital and hence inhibiting allocational efficiency can create a better environment for planning and innovation and lead to faster development than would otherwise be the case. Part of the reason for creating giant firms is precisely to afford the insulation from market pressures to allow long-term, major innovation. Arguably this point is borne out by

---

[20] Possibly one might try to assess the costs of alternative forms of social organization of production, but the controllers and owners are unlikely to opt for a form which reduces their power, even if it is superior to existing forms.

the relative performances of more liberal (allocationally efficient) economies such as Britain and the USA and more organized (dynamically efficient) ones, such as Germany and Japan.[21] The centrally planned economies also owed their rapid early growth to the ability to make major long-term investments, although critics could reasonably claim that some of them were inappropriate precisely because of the insulation from market demand. However, the fact that these different kinds of efficiency can conflict should not be construed as discounting the importance of any of them; trade-offs are to be expected in any assessment of economic behaviour, and indeed are the standard fare of mainstream economics.

In other words, these last two points show that to appreciate why allocational efficiency is less important than liberals suppose, we have to correct the liberal optic's bias towards markets and look beyond them into production, understood as the transformation of inputs into products (including services) and the development of new products by means of social labour under particular forms of organization. While radicals typically underestimate or fail to understand the importance of allocational efficiency, liberals typically overestimate it relative to other kinds of efficiency.

The Panglossian Assumption

One of the most common assumptions about competitive systems such as markets is that they encourage the survival of the fittest. This is related to the 'Panglossian assumption' that whatever eventuates under competition, must be optimal. Thus the leading firm must be the most efficient. If an industry is highly vertically integrated, it must be because this is the most efficient form of organization. The Panglossian assumption is implied not only by arguments about allocational efficiency, but by Hayek's defence of the 'extended order' that has evolved through competition. It is even evident in radical political economy, albeit in diluted form, and with the qualification that judgements of efficiency are relative to the prevailing capitalist order. According to Panglossian views, even though individual actors and organizations may not always try to optimize, the competitive system favours the survival of the fit over the unfit, so that the results approximate the optimal outcome. This is an attractive argument for defenders of market economies, for it serves

---

[21] The familiar consumer video took 14 years to develop; while European and US firms gave up the attempt to develop this product, Sony, with its relative insulation from short-term market forces, persevered and was first to offer a saleable product.

to protect claims about the optimality of market outcomes from empirical criticisms referring to the fact that economic agents often make no attempt to optimize. At the same time it should be noted that the Panglossian assumption could also be used to defend non-market kinds of competition, such as political competition or the competition for promotion within organizations.

The Panglossian assumption fits well with an *evolutionary* view of competition. The history of political economy is marked by interchanges with evolutionary biology, indeed biological analogies predate equilibrium theory, and Herbert Spencer's concept of the survival of the fittest influenced not only Darwin but economists today. However, recent theories of evolution discredit these simplistic analogies, and as Hodgson (1993) shows, the lessons for economics are far reaching (see also Nelson and Winter, 1982; Storper and Walker, 1989).

At one level the notion of the survival of the fittest seems a truism, indeed it is close to a tautology: to survive in competition an organism must be fit, and a fit organism is one which survives in competition. Nor is it limited to the case of equilibrium, for development under competition is commonly assumed to involve the selection of the fittest. Once these apparently innocuous assumptions are accepted it is but a short step to the momentous normative proposition that we should not intervene in the competitive, evolutionary process, unless for reasons which do not concern efficiency. However, the concepts of the survival of the fittest and of environmental selection lose their apparently unproblematic character immediately we scrutinize them:

1. First, the naturalistic character of the analogy is problematic in a social context, for the bad may outlive the good. Survival and goodness are not equivalent: might is not necessarily right. We may not be able to do much about the determinants of survival in nature, but the criteria for being made winner or loser in society are open to normative evaluation and can be changed.

2. Fitness makes little sense independently of a specific type of environment. The shark is superbly equipped for the ocean but not for land or air. The concept of survival of the fittest tends to direct critical attention towards individuals or particular institutions – particularly those which fail – and away from the environment (e.g. capitalism) within which they operate. But then the environment includes other species (individuals, organizations, social structures) too, so that the failure of species *A* to flourish may be a consequence of changes in other species, such as the decline of *A*'s prey. By analogy, an institution which flourishes under feudalism may not under capitalism, and one which flourishes in country *A* may wither in country *B*. It is not unreasonable, therefore, to blame the environment rather than the organism, for failure.

3. Fitness does not guarantee survival, nor does survival imply fitness. Organisms/organizations may be able to set in motion positive feedback which helps them survive but which closes off the development of other, possibly superior forms. Thus the QWERTY keyboard continues to flourish, despite its inefficiency, and the VHS video system has outcompeted the technically superior Betamax system. Organizations may combine to suppress competition or find niches which protect them from competition. Prisoners' dilemma and chicken situations may lead to sub-optimal outcomes. The success and failure of certain mutations may be only local and temporary, and promising lines of change may be foreclosed. Where there are increasing returns to scale and external economies of scale and scope, success will go to first movers rather than to the most efficient over a period of time. Change is therefore frequently path dependent, so that what survives depends on the nature of the initial state as well as competition, and geographically specific lock-in effects may restrict developments along particular lines for long periods (Storper and Walker, 1989). This evolutionary process can be likened to a trial-and-error process, but it is one in which each trial changes the environment for the next, so that whether something survives depends not only on its intrinsic qualities but on when and where it is tried out. Development should therefore not be seen as an optimal path chosen from a range of alternatives.

4. Evolution does not necessarily lead to spontaneous order, as Hayek implies (Hodgson, 1993;) piecemeal changes may precipitate structural breakdown and spontaneous *dis*order. This has devastating implications for Hayek's defence of 'the extended order'.

5. While differential rates of change in numbers of different organisms (structural change in an economy) may result from competition, such changes can also come about independently of competition, simply as a consequence of different rates of birth and death, together with migration in and out of the area in question. Selection is a product of fecundity as well as mortality. It is therefore possible for a relatively fit species which happens to have a low birth rate to be overwhelmed by a weaker, but more fecund species. Economic institutions which are efficient and can survive competition, but which are difficult to set up, may therefore remain in the minority. This appears to be the dilemma of cooperatives. The small numbers of cooperatives in contemporary economies might seem to imply that they cannot match the efficiency of capitalist firms, and yet many empirical studies of their efficiency have shown them to be the equals of capitalist firms. The paradox is easily explained by the greater difficulty of setting up collectively owned enterprises compared to starting capitalist firms (Hodgson, 1993; Elster, 1989; Estrin, 1989).

In view of these points there is no *guarantee* that economic evolution is progressive, let alone optimal. The Panglossian assumption provides an absurdly charitable view of the *status quo*. If the critique seems initially gloomy in challenging the optimality of competitive outcomes, it also gives ground for optimism, for it implies that we don't have to accept what has evolved and that there are many other possible lines of development that could be opened up and which are not necessarily inferior for not having arisen spontaneously. The critique of the Panglossian assumption is therefore of crucial importance not only for positive theory, where it challenges common narratives of the nature of change, but for normative political economy, where it undermines the case against intervention. This does not conflict with our earlier endorsement of a moderate version of Hayek's critique of constructivism, nor does it mean that we can afford to ignore the lessons of the successful or the benefits of competition. Not all successes or failures are mere accidents, and competition is still likely to moderate the power of the producer over the consumer, increase choice and encourage efficiency and innovation. In other words we have to steer a course between Panglossian views and a wholesale rejection of arguments for competition.

A less radical but convergent, anti-neoclassical line of thinking on optimality comes from Austrian and post-Keynesian sources. Under capitalism, competitive markets do not allocate resources in a static framework but provide a mechanism for changing the framework. The economic structure at any time is not necessarily the most efficient possible but simply whatever has evolved as the outcome of cumulative causation in particular areas. If individual goals are subjective, local and not collectively known or knowable, there can be no question of deciding whether the overall allocation of resources is optimal, since there is no common criterion of the good. Thus:

> the market economy, as an aggregation, neither maximises nor minimises anything. It simply allows participants to pursue that which they value, subject to the preferences and endowments of others, and within the constraints of general 'rules of the game' that allow, and provide incentives for individuals to try out new ways of doing things. There is simply no 'external' independently defined objective against which the results of market processes can be evaluated.
>
> *(Buchanan and Vanberg, 1991, p. 84)*

This lack of a clear external objective is by no means an embarrassment for liberals, for they favour systems which maximize the freedom of agents to pursue their own conceptions of the good rather than those of some collective, though this ignores prisoners' dilemma and similar problems, where individuals end up worse off than they would by cooperating or being regulated.

For Austrians, the neoclassical view that markets allocate known resources among known ends ignores the fact that this information is too diverse and complex ever to be knowable; instead they argue that the special quality of markets is that they can coordinate activities of people who are ignorant of each other's local knowledge. They therefore argue that central planning is defeated by the unavailability of this information to any single agency. But as Buchanan and Vanberg point out, this still implies that the information and knowledge already exist and hence that an omniscient and benevolent planner could coordinate everything. It thereby misses the creative aspect of markets as a setting in which people are encouraged to create new products, methods and needs which hitherto they had not thought of (Buchanan and Vanberg, 1991). Central planning is further handicapped by its inability to reproduce this local, dispersed creativity. Other institutions (e.g. groups, universities) can also promote creative change, but markets are good at encouraging and coordinating these changes within a catallaxy.

While these points are of major importance and have implications far beyond questions of allocational efficiency, it would be wrong to dismiss the latter altogether. Allocations which take into account the scarcity of resources relative to the range of competing intensities of demand for a host of uses (i.e. cost, as we have defined it) are still required. Even if we reject the view that markets optimize allocational efficiency, it has to be recognized that the performance of other modes of coordination on this criterion is almost certainly inferior. At the very least, issues of allocational efficiency deserve a prominent place on the agenda of economics and political economy. Just because most literature on the theme also ignores class and power and provides apologetics for capitalism, it does not mean that the issue is purely a bourgeois one. All economies provide ways of allocating scarce resources among competing needs, though the allocations are always relative to existing distributions of power. Removing the gross imbalances of power of class societies might reduce one source of distortion, but if markets were also abolished allocational efficiency might deteriorate further. In the case of market economies, the allocation is an unintended consequence of the struggle of enterprises to survive rather than a system goal. Given that socialists aim to produce an economy in which allocational questions become matters for decision rather than *ex post* resolution through market processes, it is all the more absurd to exclude them from political economy.

Conclusions

I have discussed five areas of debate relating to markets. In each we find problems with both radical and liberal views — in their respective dismissals and uncritical endorsements of the invisible hand and consumer sovereignty, in their selective views of the implications of markets for equality, in their respective ignorance and exaggerated enthusiasm for the allocational efficiency of markets, and in the uncritical reception given by both to the Panglossian assumption. There are both efficiency and ethical arguments for and against markets and other modes of coordination. As any acquaintance with political philosophy shows, the relationships between these types of argument are themselves exceedingly complex, as efficiency and equity considerations can sometimes be in harmony and at others in conflict (Buchanan, 1985). Unqualified approval or condemnations of a mode of coordination based on a single point for or against it are never likely to be compelling. In particular, arguments taken in isolation for or against markets score no knock-outs. At several points we have seen that not all the features generally attributed to markets are unique to them. Some are likely to be found in non-market institutions and economies too. This is an important point for comparative evalutions of economic systems.

Furthermore, we have only presented an *abstract* analysis of markets. The actual behaviour of people in concrete markets cannot be read off simply from market structure (though it is constrained and enabled by markets) and hence cannot be predicted and judged purely a priori. Since markets are always embedded to some degree and since there are always other modes of coordination present even in the most marketized society, their ideological effects may be offset by other influences. Ironically, as Hirschman points out, neoclassical defenders of markets who assume perfect competition have denied themselves the possibility of appealing to their socially integrative effect by ignoring embedding and the dense interaction needed to exchange information in real markets, presenting instead models of anomic spot markets in which people already know everything they need to know and only need to exchange commodities for money, and depart (Hirschman, 1982). By the same token, radicals can more easily present markets as anti-social by ignoring embeddedness. Moreover, as we shall see, markets can be coupled with different forms of ownership, and these also have important consequences which can be easily confused with those of markets *per se*. The same goes for other modes and settings. Much depends too on the particular nature of the products or services concerned. An arrangement which is harmless for the production and distribution of cakes may be highly damaging and divisive for the delivery of health services. No single argument for or against a particular mode of coordination can trump all the rest: trade-offs and compromises are unavoidable.

6

Ownership and Control

The identification of capitalism with the market has long been, and continues to be, a ploy of capitalist apologetics to distract attention from other essential characteristics of capitalism that are far more problematic, notably private ownership of the means of production and, above all, wage labour.

(Schweikart, 1992, pp. 29–30)

It is the failure of both the political right and the left to disentangle the concepts of private ownership and the competitive market that has led to the premature obituaries for socialism.

(Bardhan and Roemer, 1992, p. 102)

In chapter 3, we saw how the distribution of power in capitalism depended not only on 'vertical' relationships of class, in terms of ownership and control of the means of production, but on 'horizontal' relationships across the division of labour. Morever, we identified a dialectic between social relations and the material characteristics of economic activities. Issues of ownership and control depend fundamentally on this dialectic. Ownership is a key source of economic power but it should already be clear that the capacity to influence and control economic activities can derive from several other sources too. Owners of means of production in particular can have considerable influence on the shape of the economy, but their powers are limited not only by the actions of workers, consumers and the state, but by the material and informational constraints of the activities and resources they control. Since economic activities are not infinitely plastic, and differ considerably in their material and informational properties, the various forms of ownership yield different degrees and kinds of power according to the kinds of objects or services concerned.

While ownership is a political economic matter of great importance, debates about it tend to be rather simplistic. Markets and private property are often taken as self-evidently synonymous. The standard state-versus-market or public-versus-private frameworks that dominate political economic debate fail to do justice not only to modes of coordination but to these forms of ownership and control. In underestimating the variety and possible combinations of modes of coordination and ownership forms, they also underestimate the range of organizational types and behaviours that can be expected in economies, thereby impoverishing political debate. Voluntaristic thinking about ownership, in which material and informational dimensions are ignored, is particularly common in radical advocacy of social ownership.

In this chapter I shall examine ownership and control in more depth, with the aim of developing a clearer idea of their relationship to modes of coordination, and of their material and informational limits. This is not merely an exercise in analytics, for it also sketches out the practical potentialities and limits of the various forms of ownership. I shall begin by counterposing Weberian, Marxist and Hayekian views of property in means of production, and negotiating between some common radical and liberal views of private property. I shall then compare the qualities of social ownership, and move on to two often-neglected kinds of ownership – majority or universal private property, and restricted collective ownership. Next, workers' 'self-ownership' of their own labour is discussed as an important issue affecting judgements of who is entitled to what. I then discuss externalities as a situation where property rights are not, or cannot, be defined and relate these to the divisions of labour and knowledge. The chapter concludes with some comments on property and responsibility and the wide range of concrete behaviours associated with different forms of ownership.

Private Property

Property is not merely a relation between individuals and things but a social relation; in claiming an object as my own private property, I am doing so over and against possible rival claims of others and defining what I can do with it in relation to them. Such claims to formal ownership are normally backed by the legal authority and sanctions of the state. While ownership is normally assumed to give or confirm control, it does not always do so, and sometimes it may be possible to have significant control or influence over property without ownership (Murray, 1987). Firms with monopsony-power, such as supermarket chains, may be able to exercise substantial control over suppliers without owning them. Since ownership implies responsibilities as well as rights, and since market relations do not involve obligations other than those of contract, this kind of control through exchange relations without ownership may be preferable. Whether the title of ownership yields any power depends on how valuable the object is to others and the extent to which its ownership limits what they can do. Minority ownership of the means of production is the paradigm case of this, though here the resulting relations are dialectical rather than absolute; the owners need propertyless workers if their property is to be a source of power.

Ownership is not absolute but covers varying bundles of rights and obligations regarding the use and disposal of property. On the whole, the rights of owners to use and dispose of property increase with the development of markets, while the obligations to others decrease. The decline of usufruct (whereby, for example, non-owners of land might have certain grazing rights on the land of others) made an important contribution to proletarianization (Polanyi, 1944; Landes, 1969). However, private ownership has still not been absolutized totally and is invariably qualified to some degree. In many cases it may be more accurate to say that people own rights relating to resources rather than simply that they own resources (Steele, 1992). Thus land ownership rights are normally qualified and regulated by planning and other laws. It is not merely markets which are regulated but the use of property, including private property, outside the moment of exchange. These constraints on use can be as mundane as the rules of the Highway Code, and they regulate a vast range of activities. Yet they are systematically ignored in liberal theory, which works with a metaphor of sovereign individuals surrounded by fences enclosing their private sphere and property, and into which others – particularly the state – have no right to encroach (Haworth, 1994). Against this influential metaphor it bears noting that ownership is not an all-or-nothing matter based on an absolutely clear-cut distinction between private and public. There are therefore plenty of important precedents for policies which limit or define private property rights in particular ways, without abolishing them altogether.

Whether something can be owned and how far it can be controlled depends partly on its material qualities. Objects vary in their divisibility, reproducibility and cost, and hence their suitability for different kinds of ownership; what is suitable for socks is not for jet aeroplanes or steel works or scientific ideas. As Harvey (1993) points out, the structure of the environment is often not sufficiently fixed or clear for it to be possible to establish property rights; it would be difficult, for example, to establish private property in shoals of fish in the ocean or air-borne pollutants. Similar problems occur regarding the ownership of information.[1] Moreover, the scale and divisibility of economic activities affect

---

[1] See Lury, 1993 for a discussion of how the rise of electronic media has influenced property in cultural products.

the nature and degree of control that individuals can have. As Weber argued, 'the expropriation of individual workers from ownership of the means of production is in part determined by purely technical factors' (1947, pp. 226–7).

> As long as there are mines, furnaces, railways, factories and machines, they will never be the property of an individual or of several individual workers in the sense in which the materials of a medieval craft were the property of one guild-master of a local trade association or guild. That is out of the question because of the nature of present-day technology.
>
> *(1972, p. 199)*

Wherever Weber discusses this issue he is always careful to include the qualifier 'in part', implying that he does not deny that workers were expropriated by other means. Moreover, he notes that these technical factors 'do not exclude the possibility of expropriation by an organised group of workers, a producers' co-operative' (1947, p. 227). However, he seems not to regard this as significant, for he also claims that it makes no difference whether the organization in which the worker is employed is headed by a capitalist or a minister of state – or, presumably, a council of workers. I shall argue that this latter claim is wrong, partly because control and ownership are confused in the first of the above quotations: certainly the scale of production technology affects individual workers' control, but so too do relations of ownership.

The most familiar Marxist views of ownership are concerned with capitalist private property and class and are counterposed to the Weberian, technical explanation of the divorce of workers from the means of production. Capitalist private property in means of production is different from other forms of private property in two respects. Firstly, it can only exist on a minority basis; if everyone owned means of production, owners could not be capitalists. Pots and pans are a means of production of meals, but if everyone owns them, no one derives any power over others from this. Secondly, the extraordinary thing about capitalist ownership is not merely that it involves the means of production, but that it extends to the product of workers' labours.[2] If I buy shoes from an independent, self-employed producer, the actual producer – the one whose labour has been 'mixed' with the raw materials – owns them until they are sold. If I own a shoe-making firm, I get to own the products of my employees' labour from the start. This second aspect is often overlooked in Marxist theory, but it supports the key Marxist argument that wages are a payment for workers' labour power rather than for

---

[2] Under special conditions, enterprises may hire out means of production to others, but usually there is more profit in retaining control over it and hiring workers.

their labour. If I expect a profit, the wages I pay the workers cannot be equal to the value of what they produce, even allowing for the costs of replacing worn out and used up means of production. The denial of the common idea that wages are a payment for work done is not merely a theoretical matter, but is reinforced in law through the recognition of the product of the employee as belonging to the employer, not the employee.

However, while one normally associates Marxist analyses of property with these kinds of views of class and ownership of the means of production, in Marx's early work, property is far more strongly related to division of labour than class.[3] In *The German Ideology* Marx and Engels describe division of labour and private property as 'identical expressions: in the one the same thing as is affirmed with reference to activity is affirmed in the other with reference to the product of the activity' (1974, p. 44). 'The various stages of the division of labour' are viewed as 'just so many different forms of ownership. . . . [W]ith the division of labour . . . is given simultaneously the distribution, and indeed the unequal distribution, both quantitative and qualitative, of labour and its products, hence property' (p. 33).

These might seem rather surprising statements. How far are they true? In pre-capitalist societies property is certainly less well-defined in relation to individuals (D. Sayer, 1987, p. 60). Where there is little division of labour and many tasks are shared, particularly ones that only have to be done intermittently (e.g. harvesting), the concept of private property has little relevance. But where there are separate, largely self-sufficient producers, there clearly can be private property – chiefly in land, either with owner-occupiers or landlords without division of labour. However, the more individuals or groups come to specialize in particular activities, the more they need to be assured of *de facto* control over the specialist means of production required for their work, and over the disposal of the product. In this respect, Marx and Engels' claims make more sense. This *de facto* control is reinforced by legal titles of ownership in more marketized societies, for this stabilizes the framework of production with a social division of labour. If producers fear that their tools, products and long-term investments will be arbitrarily seized by others they will be discouraged from producing (Riker and Weimar, 1993). Hence a major reason why so little can be produced privately in the former state-socialist countries is that new property rights have not been finalized.[4]

---

[3] This is consistent with the much-overlooked fact that division of labour overshadows class in the early Marx (Rattansi, 1982).
[4] An interesting corollary of this is that in very liberal capitalist regimes, the vulnerability of firms to takeovers discourages long-term investment.

As we saw in chapter 3, divisions of knowledge and labour themselves give specialist producers a degree of power, even if they do not own the means of production, simply because outsiders lacking that know-how are dependent on them. Just how much power they have depends on the nature of their work and how it impacts on other people – whether they are electricity supply workers or university lecturers for example – though the resulting distribution of social power can be modified by strategies of closure and discrimination used by workers against other workers and consumers. Under capitalism, access to the power that workers derive from doing particular kinds of specialized work is in the gift of their employers, but remains none the less. This source of control does not depend on formal ownership; it exists in any economy and may in fact be counterposed to class power.

If we are to understand the 'vertical dimension' of production in relation to ownership, it is helpful to refer again to Marx's 'double-nature' thesis. Recall that the hierarchical organization of capitalist firms was, for Marx, partly a consequence of the technical necessity of coordinating horizontal divisions of labour within organizations by vertical divisions of labour, and partly a consequence of the capitalist ownership of the means of production, the latter being a dispensable feature of large-scale production. Capitalists may of course do more than merely own the means of production, and actually participate in the technically necessary organizational work. But here, we can still maintain a distinction between work that would be technically required in the absence of capitalist ownership and work that involves enforcing the powers they derive from their position as owners (Marx, 1972, ch. 23). Many authors have failed to grasp the double-nature argument by conflating technically necessary hierarchy with contingent forms of ownership. Thus, contrary to the claims of writers such as Samuelson (1989) and Arnold (1989), while workers in a cooperative are likely to be driven by competitive pressures to set up a hierarchy and appoint a manager they do not have to place their futures in the hands of capitalist or quasi-capitalist entrepreneurs and give up ownership of their products to them (Schweikart, 1987). As Marx put it: 'In a co-operative factory the antagonistic nature of the labour of supervision disappears, because the manager is paid by the labourers instead of representing capital counterposed to them' (1972, p. 387). (However, other, non-class forms of antagonism are likely to persist.) The property relations central to capitalism have no technical warrant in large-scale production. As Marx observed, although a conductor is needed to coordinate an orchestra, it is not technically necessary that he or she owns the instruments and pays the musicians' wages, or has exclusive rights over earnings from concerts and recordings (to up-date the example).

From an entirely different perspective, Hayek provides another interesting but flawed attempt to link property with the divisions of labour and knowledge. As we saw in chapter 3, he argues that private property is

> 'generally beneficial in that it transfers the guidance of production from the hands of a few individuals who, whatever they may pretend, have limited knowledge [i.e. state planners], to a process, the extended order, that makes maximum use of the knowledge of all'.
>
> *(1988, pp. 77–8)*

Here Hayek relates private property to information and the division of knowledge; markets and private property are supposed to allow power, responsibility and specialized knowledge to correspond, whereas central planning divorces power from inherently localized and specific knowledge and information. While Marxists see property primarily in the vertical dimension of control over some specific process, Hayek sees property largely with regard to the horizontal dimension: private property gives actors the powers and responsibilities necessary for responding to unknown others through the market, for coordinating the social division of labour. It allows individuals (including enterprises and other institutions) freedom to pursue their own ends without needing to account for them to others, makes them responsible for their choices and inhibits free-riding (Holton, 1992, p. 59). These are important qualities of private or fragmented ownership and control which any political economy of advanced economies must recognize.

While private property makes individuals responsible for their actions relating to that property, it only encourages them to be responsible in relation to their own self-interests and in terms of spending carefully to save money. It does not rule out anti-social or environmentally unsound actions. Division of labour coupled with the institution of private property can divide up responsibilities to a certain extent, but it cannot abolish all the continuities and interdependencies among social activities or between these and nature; indeed, the higher the level of division of labour and economic development, the denser and wider the interdependencies. The division of responsibilities and control resulting from fragmented ownership of activities which are highly interdependent can be economically irrational, as in the case of having separate companies owning different parts of a national railway system. On the other hand, as we shall see, social ownership and the abolition of markets does not always make it easier to cope with these interdependencies.

Yet even if we accept this as a qualification rather than a negation of Hayek's point about private property and responsibility, there are still further problems. The main one is the deployment of the characteristics of catallaxy to legitimize not merely clearly defined ownership or rights of use by separate individuals or bodies, but capitalist private property in the means of production and the employment relation, as if capitalists and propertyless wage labourers were needed just because knowledge is dispersed. In so far as it allows the powers deriving from formal ownership and specialism to coincide, fragmented ownership coupled with markets and networks might generally encourage the fullest use of the specialized knowledge that is dispersed across any advanced division of labour. But the enterprises don't have to be owned by capitalists; they could be owned by self-employed petty producers or by workers. Here, as in the work of many economic liberals, the exchange optic, with its focus on horizontal relationships, serves to distract us from the vertical social relations of production. The rough equality of individuals exchanging their products in (pre-industrial?) markets is used firstly as a cover for the relations of domination of capital over propertyless labour and secondly as an evasion of the fact that some of the 'individuals' in advanced economies are large, powerful organizations.

These are significant deceptions, but it is important to realize both that they are independent of one another and that they do not undermine Hayek's central claim here. They are independent because large organizations do not need to have capitalist social relations of production, so that drawing attention to the imbalance in size and power of 'individuals' is not getting at specifically capitalist characteristics. They do not undermine Hayek's central claim because he could reply that whatever the legitimacy or otherwise of capitalist ownership, it at least shares with other forms of private property the virtue of allowing a rough correspondence between power, local knowledge or know-how and responsibility within a catallaxy or advanced social division of labour. Similarly, large, independent organizations could be justified on the grounds that they correspond to specialisms and local knowledges involving large numbers of people. However, like many liberals, Hayek generally avoids talking about large organizations, preferring not only an atomized view of society as an aggregate of individuals, but a similarly atomized view of knowledge and information. He therefore tends to assume that economically relevant knowledge is mostly highly specialized and localized. Yet, as we saw earlier, some economic knowledge has to be common and shared across wide constituencies. Ironically for Hayek, who wants to reject the concept of the social, the latter includes knowledge of markets and business law and procedures (Hayek, 1988).

Further objections could be made to Hayek's treatment of the relationships between ownership, division of labour, information and know-how. Individuals, groups and institutions do not just happen to have specialized knowledge; they actively protect and attempt to monopolize it, for it is usually to their advantage to do so. The distribution of local, specialized knowledge and information does not pre-date competitive markets and private ownership, but develops with them. Where institutions and individuals have to compete to survive this generates well-known problems regarding the ownership and exchange of information, with each individual or institution reluctant to share their own secrets but eager to benefit from others'. In these circumstances, which are not unique to *market* competition, the pursuit of self-interest may lead to collectively sub-optimal outcomes through failure to share helpful knowledge with others. This is not to suppose that in the absence of such strategic behaviour the division of knowledge would disappear; specialisms would still be largely opaque to outsiders, but in any division of labour, access to information is likely to be contested. The distribution of specialist knowledge and information is neither purely a technical consequence of division of labour or catallaxy, nor purely a result of power struggles; it is always both, with power being developed partly through and in response to the division of labour and knowledge, and the latter responding in some degree to those power struggles. This is just another instance of the dialectic of forces and relations of production noted in chapter 3.

The power of private producers in a division of labour coordinated by markets therefore derives from several sources, from their exclusive ownership of valuable means of production, from their specialist knowledge, from employing wage labour – if they do – and from the strategies they employ, including those of closure and secrecy against rivals. Each of these feeds into and can reinforce the others, such that it is all too easy to mistake the effect of one for the effect of another. Hayek provides an incomplete and tendentious view of the relationships between these sources. Ownership of means of production does not follow automatically from the possession of local, specialized information and expertise – it has to be won through primitive accumulation or through expropriation and exploitation, or bought from others. But what of the converse relation? Does local knowledge follow from the ownership of property? To some extent it does; ownership of means of production makes it easier to develop specialist knowledge, and it gives one specialized needs. Hence, if I acquire the equipment for making a good, such as shoes, then I soon develop a set of specialized needs relating to that equipment and its use and maintenance, and since I know better than the government or distant others what these needs are, I can then use Hayek's argument to legitimize my ownership of that equipment. Yet while Hayek provides a justification for the existence of fragmented or private ownership, and one to which I have given qualified support, it says nothing about *who* should be the owner, or how they should acquire that ownership, or the particular form it takes. It makes no difference to his argument whether the private property was accumulated by individuals through their own labour or by individuals expropriating what was formerly socially owned or unowned. Both from the positive point of view of explaining how private property develops and what its effects are, and from the normative perspective of evaluating such arrangements, we need to separate out these aspects instead of explaining and justifying one by reducing it to or hiding it behind another.

These flaws expose the apologetic elements of Hayek's thesis regarding capitalism, but they also protect it from the charge that it is nothing more than capitalist apologetics, since the same argument could be used to defend a system of petty commodity or self-employed producers or even our ideal type of market socialist system. In sum, then, there is no question of plumping either for the Marxist vertical view of property or the liberal horizontal view. Power and control in an advanced economy relating to property ownership involve *both* vertical and horizontal dimensions. Neither the one-sided view of Marxism nor that of liberalism will do.

Private ownership is often discussed as if owners had to be individuals, but institutions can be legal owners, as in the case of joint-stock companies. Their development has been widely interpreted as involving a separation of ownership and control between directors and shareholders, and in some cases, as marking a fundamental shift in the character of capitalism. However, this development certainly does not mean that ownership has become divorced from the company; in fact ownership becomes identified with the company itself rather than with individuals. The directors act as trustees or agents of the institution. They have control and responsibility, but not ownership. In Britain, shareholders do not have any ownership rights in company assets, which belong to the company as an independent legal entity, though they can appoint managers and directors – which may include individuals who are also shareholders. 'Stockholders usually act as investors or *rentiers* of capital rather than as the owners of corporations' (Abercrombie et al., 1986, p. 133). In effect they have a degree of indirect influence over the companies in which they have shares through the rights to revenue, but they do not have ownership, control of the physical assets or responsibility for them, as this belongs to the company itself, though large shareholders may be able to wrest control in the name of the company (Tomlinson, 1982, pp. 65, 82ff). Shareholders have minimal barriers to entry and exit, but only limited and unattractive options for voice, in terms of changing company policy. Dividends are at the discretion of the board of directors, but where companies rely extensively upon equity they are likely to be paid out under most circumstances, quite apart from their own stock interests, the directors are under pressure to pay out dividends simply to stop the value of the shares falling and making the company vulnerable to takeover. Under these circumstances, long-term investment is difficult and the quarterly returns rule the day. Shareholding is generally dominated by a few large individual and institutional holders, so the dispersion of control associated with the depersonalization of ownership resulting from the growth of stock ownership has been limited.[5] These developments make little difference to the capital–labour relation: workers still produce surplus value, the directors carry out the functions of capital and they and their creditors and shareholders appropriate much of the surplus value.

Other forms of raising credit also disperse control away from the enterprise itself towards creditors who naturally want to ensure their loans are repaid. Whereas exit is generally a more attractive option than voice for dissatisfied shareholders, creditors such as banks have more interest in exercising voice. The relative influence of shareholders and banks varies across countries, with implications for the security and stability of enterprises. Many major Japanese firms belong to industrial groups in which members have reciprocal stockholdings, and which have their own banks, both supporting and depending on the group's companies. These groups not only offer mutual support and protection from takeover, but share know-how and police one another in order to ensure growth. In these circumstances, the rights of key workers and managers may take precedence over the rights of shareholders in times of difficulty (Dore, 1983: Sayer and Walker, 1992). The implication of these variations is that private ownership is not absolute or restricted to a single form. There are many forms and the differences between them can have a significant effect on behaviour, particularly regarding the security of companies and their scope for long-term initiatives.

This dispersion of interests in and influence over enterprises is not a uniquely capitalist phenomenon, nor is it necessarily undesirable. Under any economic system the growth of enterprises would be severely and unevenly restricted if they had to rely on raising capital internally. In an economy of worker-owned commodity-producing enterprises, some kind of dispersion of control and possibly ownership beyond the enterprise would also be needed, though it could be under some form of public

[5] While several million people bought shares in privatizations in Britain, many immediately sold them to centralized financial capital, and those that didn't gained no effective influence over the industries.

control, as in the case of the Mondragon cooperatives (Thomas and Logan, 1982).[6]

Social Ownership

It is common to counterpose capitalist private property in the means of production (which Hayek wants to legitimize, albeit in disguise), to social ownership, in which every member of a whole society owns property in common. In an advanced society, 'social ownership' is largely meaningless (Aganbegyan, 1988; Brus and Laski, 1989). For material and informational reasons to do with division of labour and knowledge and spatial separation, it is usually impossible for everyone to have some recognizable and roughly equal influence over what happens to property. Attempts at achieving social ownership on a large scale therefore tend to resolve into centralization and delegation. In the former case, the state acts ostensibly on behalf of the people, controlling the use of property in broad outline, though without owning it. Thus, actual control of economic activities under state socialism derives not from legal ownership but from position within the social division of labour and the hierarchy that attempts to direct it. Instead of equalizing power, social ownership therefore tends to concentrate it in the state, which then has power over those who are nominally co-owners. At the same time, the state cannot overcome the divisions of labour and knowledge and change the catallaxy into an economy*. This is why social ownership is often restricted to particular activities or sectors (i.e. economies*), and why, even within these, the state has to delegate power to specialist managers. Managers of nationalized industries and other public institutions therefore become like property owners to the extent that they have the right to deploy resources as they see fit (Steele, 1992).

Even where a service might need to be provided centrally for everyone, the work involved in providing it is often highly specialized and not easily made the subject of democratic control. As a user of a local park, my rights are equal to anyone else's, but I cannot expect to have the same kind of control over its development and maintenance as the park's staff. Although nominally any specific means of production belongs not to those who actually work with it but to the whole of society, it is of course only the former who are actually empowered to use them: things

---

[6] See Bardhan and Roemer, 1992 for proposals for a market socialism involving both dispersed ownership and control and industrial groups on lines influenced by Japanese examples.

could hardly be otherwise in an advanced economy. Again, as under capitalism, workers derive a certain degree of power over others simply by virtue of their attachment to particular jobs and means of production. Neither social ownership nor central planning can override this power completely. Similarly, state housing is rented out to particular households who have certain rights over the use of 'their' houses. This may seem to be stating the obvious, but it shows that in practice social ownership in advanced economies cannot mean control of everything by everyone.

In practice, what is termed 'social ownership' invariably takes the form of 'public ownership', where control is delegated to specialist managers. Their control over the technical divisions of labour in the services in question and the lack of competition may give them considerable powers, rivalling those of capitalists, but whereas capitalists are not needed to run specialist activities, specialists are, and it is from this that the roles and powers of managers and professionals in public services gain their legitimacy. When extended to a whole economy, public ownership and social ownership correspond respectively to state socialism and infeasible socialism as located in figure 3.3.

For liberals, social ownership lacks a means of regulating use or attributing responsibility, and it subjects individuals to the will of the community or state, which, liberals assume, will have less interest in and knowledge of their predicament. As the fable of the 'trajedy of the commons' claims to show, social ownership is a recipe for free-riding, neglect of resources and failure to ration their use (Hardin, 1968). If there are a hundred users sharing common pasture, an individual grazing an extra animal will be able to appropriate all the gains, but will suffer only a hundredth of the losses caused by their actions. Others will be tempted to follow suit, leading to the inevitable degradation of the pasture. One solution is to create private-property rights in the pasture, so that responsibility is clearly defined; the other is to regulate social use. However, as has often been pointed out, the use of many commons is already regulated – and hence not susceptible to the tendencies identified in the fable. Nor does private ownership necessarily solve the problem, for it is not always in the interest of private owners to conserve resources. Thus, privatization of the Amazon rainforest has accelerated its destruction. A smallholder might have an interest in sustainable exploitation, though market conditions might prevent them achieving this, but an international logging company has little interest in conserving trees in particular areas even if it owns the land on which they grow. The Right can respond to such points by arguing for fuller and better property rights, but often this is infeasible. More importantly, taken to its logical conclusion, the attempt to extend private-property rights leads to an erosion of the most important kind of commons, one on which social life depends – trust. As Haworth points out, the great irony of the Right's combination of enthusiasm for private property and abomination of the state, is that private property needs the backing of the state to exist, and a legal system to police violations. The greater the privatization, the greater the scope for litigation and the role of the state's judicial functions (Haworth, 1994).

Social ownership is not necessarily a free-for-all. In part the liberal resistance to social ownership derives from its attachment to individualistic behavioural assumptions, according to which (rational) individuals act only in accordance with their own narrow self-interest, and will not miss a chance to free-ride or engage in beggar-thy-neighbour behaviour, and from its abstraction from the rules and rights of use attaching to social property. Frequently, these assumptions are defended by the argument that micro-theory needs micro-foundations. This is reasonable (provided it is not assumed that the whole can never be more than the parts), but what is curious is that micro-foundations are so often assumed to involve asocial individuals,[7] rather than ones whose identity, norms, habits and interests are partly bound up with those of others, and whose use of property, whether private or social, is regulated. However, while individual behaviour need not be narrowly self-interested, and hence the behavioural assumptions of rational-choice and public-choice theory are suspect, the irony is that implementation of policies informed by these theories corrodes actors' sense of social responsibility and encourages them to act selfishly. The individual of rational-choice theory is more likely to indulge in free-riding than the socialized individual brought up in a context where communal responsibility and social ownership are respected. On the other hand, the problem for supporters of social ownership is that there are often good reasons for not respecting it, and these are particularly likely to be found in advanced economies – (a) because of their size and the diversity of interests within them; and (b) because the state control needed to regulate social property creates a force over and against the people.

Socialists often favour social ownership in the belief that it replaces competition (bad) with its opposite, cooperation (good). Aside from the question of whether or how far competition is bad, as we saw in chapter 3, this overlooks the vital point that the opposite of competition is also monopoly. Instead of an antagonism between competing (sets of) producers which is frequently to the advantage of consumers, one has a strengthening of producer interests over consumers. Private monopolies

---

[7] See O'Neill, 1992 for a critique of the treatment of individual and altruistic action in economic philosophy.

can of course arise spontaneously out of competion, and radicals often justify public ownership on the grounds that it safeguards the public interest by ensuring that 'natural monopolies' are not under minority private control. Yet while public monopolies deny power to private capital, they are not necessarily any more responsive than private monopolies to users or consumers.[8] In Britain in the eighties, the Conservative government cleverly exposed the dangers of 'social ownership' by presenting the sale of public assets as 'giving industry back to the people', using a socialist slogan against the Left. Although this was entirely disingenuous, the fact that they even considered this rhetorical strategy worth trying suggests that they knew many people were dissatisfied with public provision and did not associate it with popular control. Producer interests can be organized and developed more easily in monopolies than where producers are working for separate, competing private owners dependent on consumers who have the possibility of exit if they are dissatisfied.

Publicly owned industry and services can be operated in several different ways. At one extreme, which can be termed 'state capitalism', they can be run according to capitalist principles, depending on revenue rather than tax-financed budgets, raising capital on the private market and competing with private capitalist firms.[9] At the other extreme it may be budget financed and run as a monopoly public service. They also vary according to whether they are run as a single monopoly for a whole country or as local monopolies, with corresponding inplications for decentralization of political control. There can be competition within the public sector, as in the higher education system, and budgetary squeezes to encourage productivity increases, though particularly with services like health care this often means a deterioration in the service. Nationalization changes a few things, as does privatization of state enterprises, but the net changes depend on far more than ownership. Just as one cannot read off, in the manner of the 'structure–conduct–performance paradigm', the behaviour of individuals and firms in markets from market structure, nor can one read off the behaviour of public organizations from the mere fact of public ownership (Auerbach, 1989; Burawoy and Lukacs, 1989; Maynard and Williams, 1984; Sayer,

---

[8] Neo-liberals tend to dispute that natural monopolies exist, or argue that even if they do, an element of competition can be introduced by getting private firms to compete to run the monopoly for a specified period under franchise or subcontract arrangements. Whatever the practical performance of such arrangements, they illustrate how both conventional and radical thinking had underestimated the possible combinations of ownership and mode of coordination.

[9] State capitalism of this kind is quite different from the so-called state capitalism of the former Soviet bloc, which did not operate within a competitive market context.

1992b). How far they operate in the public interest and how accountable they are, how responsive they are to voice, depends on how they are run; hence popular dissatisfaction with public services, which Thatcherism exploited so successfully in Britain, is not found in all advanced capitalist countries. As Tomlinson argues, far too much credence has been given on the Left to the assumption that nationalization necessarily changes everything (Tomlinson, 1982). The same goes for those on the Right regarding the impact of privatizations.

Majority or Universal Private Property

Radical critiques tend to assume that 'private property', especially in the means of production, must be held only by a minority, as in the case of capitalist social relations of production. Yet there can also be universal or majority private property, as in petty commodity production, in which property ownership is roughly equalized. The same distinction applies to goods used for consumption too; they may involve a minority (second homes), or everyone (clothes). This means that contrary to popular assumption on the Left, private property cannot be taken as a byword for inequality – only in particular cases is it associated with inequality. Critics of private property frequently overlook this majority kind (or dismiss it as 'petit bourgeois', as if that constituted an argument), and wrongly suppose that the only alternative to the minority kind is social or public ownership. But a further option is to equalize property ownership by extending it from the minority to the majority.[10] Radicals may still object to private property for other reasons (such as its alleged anti-social effects even where it is held universally), but if they are to criticize it they need to be clear about the grounds on which they oppose it.

Restricted Collective Ownership

Between the extremes of individual private property, whether held by a majority of individuals or a minority, and social ownership, lie cases such as cooperatives, where a group having some special involvement in an activity, jointly owns the means of production it uses, as in our ideal

---

[10] The difficulty that British socialists have in recognizing this 'third way' was evident in the debates about Conservative policies for increasing home ownership in the eighties, though given the level of house prices in relation to the lowest incomes it is highly unlikely that home ownership could become as universal as the ownership of clothes.

type of market socialism. We shall term this 'restricted collective ownership', though 'private group ownership' would serve equally well, for such an arrangement meets the criteria that Hayek invokes in his defence of private property. Naturally the power of any individual worker-owner is diluted as the organization enlarges, but it is still related to a fairly coherent and manageable range of activities and responsibilities in which they are heavily committed through their work, rather than to a disparate and distant set of activities in which they are only marginally involved. This correspondence between involvement in production and ownership also has its drawbacks. As the Webbs pointed out, it would create substantial producer power over and against the interests of non-producers if generalized (Tomlinson, 1982). Though more democratic than capitalism it does not empower those who are unable to get paid work. Partly socializing ownership in larger institutions responsible for coordinating and providing finance for them would spread power, but the interests served would still be labourist and producer oriented.

Self-Ownership and Entitlements to Property

This is an appropriate place to comment on the common assumption of 'self-ownership' – in particular the view that individual people are the sole rightful owners of the fruits of their labour. This assumption is implicit in theories as different at Nozick's libertarianism and Marx's theory of value (Nozick, 1974; Kymlicka, 1990). Self-ownership might seem entirely proper from the perspective of workers, but it presents problems for those who are unable to work. While the concern of Marx's theory is to expose the source of profit, rent and interest in surplus value appropriated by capitalists, landlords and rentiers, it ignores the legitimate claims of non-workers such as children and the infirm, leaving no room for any allocation to them. Similarly with unpaid domestic and caring labour; though it is largely done under patriarchal domination and appropriated by men, it is also done for others who are unable to do it themselves, such as the young and elderly. In both cases, the element of appropriation can be recognized without assuming that workers – paid or unpaid – have exclusive rights to the fruits of their labour. (I leave aside the issue of whether housework should be waged.) The rights of the young, the elderly or the infirm can only be recognized by accepting that producers do not work purely for themselves. We have already noted how certain forms of economic organization favour producer power at the expense of others, but if we are to support the interests of non-producers in influencing the direction of economic development we have to question the assumption of 'self-ownership' itself. When we talk of 'social labour', we therefore should mean not only that what is produced is for use by others because of the existence of a division of labour, which tends to mean only those others who can pay for the products, unless subsidized by the state. There is also another sense of 'social labour' which refers to work undertaken to provide public goods or to support individuals who, for reasons of age, infirmity or whatever, cannot work for themselves. Clearly such labour does not presuppose division of labour or a cash economy. Just who should support such projects and individuals is of course a moot point (kin, community, taxpayers?) But it would be extraordinary to deny those who cannot work any rights to what is produced, as the assumption of absolute self-ownership implies.

This raises the important question of who gets property rights and who is entitled to them. One might accept the legitimacy of private ownership or restricted collective ownership in principle, but dispute who is entitled to it. To accept that private property has certain advantages is not to legitimize the existing pattern of private-property ownership. Socialists often talk as if everything should be considered as social or common property, as in the metaphor of the social cake which needs dividing up and distributing.[11] Yet a socialist who bakes a cake for his or her family is not likely to take kindly to outsiders claiming part of it as theirs. Some property is privately owned by entitlement through being the product of its producer's labour. As Nozick argues, property comes into the world with entitlements attached to it, though as we have seen, the self-ownership assumed here can hardly be taken as absolute (Nozick, 1974).

Externalities

We noted earlier that the possibility of establishing property rights depends partly on the material properties of the objects concerned. Economic externalities and free-rider problems, in which economic activities have positive or negative effects on others for which they pay or receive nothing in return, represent cases where property rights are ill-defined or absent. In some cases it may be possible to eliminate externalities by regulation; for example, open-cast mining companies could be made to bear the costs of restoring land to its former state. It is

[11] This is ironic, for the labour theory of value depends on the assumption that the worker's labour is her own. If it were not, it would be hard to see how the capitalist could be said to exploit her.

common on the Left to regard externalities as a consequence of private ownership; the socialist alternative, more or less explicit, is to remove them through some form of social ownership. But it is not only capitalist firms which create externalities. In a competitive context cooperatives also have no incentive to internalize negative externalities, for to do so would raise costs. The presence or absence of a class relation within an economic organization is largely irrelevant to the production of externalities; what matters is the degree of correspondence between the organization's material effects and its liabilities. Social ownership could actually exacerbate negative externalities by reducing the need for specific individuals or organizations to take responsibility for the consequences of their actions. (Recall the appalling environmental record of the former Communist bloc.) Scale is also important too; one does not have to assume an undersocialized or Hobbesian conception of individual action to recognize that it is easier to enforce norms against free-riding in a small organization than a large one. Private ownership may increase individualism and selfishness but social ownership does not necessarily discourage it. What happens depends not merely on formal ownership but on the effectiveness of systems of social values and norms in influencing individual behaviour.

Blaming private ownership for externalities can also imply a voluntarist diagnosis, for it ignores the material dimension of the diverse economic activities or reduces them to the way in which they are socially organized. It assumes that externalities can always be eliminated by creating organizations, such as the state, which straddle wider segments of the division of labour and therefore internalize them. However, when the state does try to internalize externalities by taking responsibility for what was formerly carried out by separate organizations, the problems are not necessarily removed. Instead they often reappear between the divisions within the state; the actions of one department continue to create externalities for another part. Such is the intractability of the diverse interests and practices associated with a complex social division of labour, that no matter how many liaison officers are appointed and how many organizational changes are made, the problems frequently remain (Nove, 1983). Similar effects can happen with other large organizations. I am not arguing that externalities are determined entirely independently of the particular forms of organization and performance criteria in use (which would constitute a deterministic fallacy, merely inverting the voluntarist fallacy), but again that materials, people and information are not completely plastic and equally susceptible to any form of organization. The obstacles to the 'socialization' of control and power in the economy are not only current forms of private ownership but the nature of what has to be controlled.

Conclusions

Both liberals and Marxists tend to assume too close a link between private property, especially minority private property, and markets. This allows them to (mis)attribute the qualities of one – be they good or bad – to the others. However, the qualities of private property are partly independent of markets. Since private-property rights cover not only rights of sale but rights of use, markets, which are merely a setting for exchange, cannot take all the credit for the benefits of private rights of use. Moreover markets are not the only source of private property – people may also acquire it outside markets, through their own labours or through inheritance or gifts.

Meaningful social ownership in which all members of a community can have significant control and influence is only likely to be possible for limited ranges of relatively simple kinds of property and activity. In any advanced economy, property relations or at least rights over the use of resources need to be established. Social ownership in the sense of ownership of everything by everyone is meaningless; in practice rights of use are usually clearly defined and controlled, and some state property, like state housing, is specified for private use. Even in the absence of market coordination, as in state socialism, clearly defined rights over the use of many resources are needed. Market and non-market exchange (as defined in chapter 4) require not only clearly defined rights of use but ownership; to be able to sell something it must be yours to sell, or you must be empowered to act on behalf of the body which owns the property. Particular kinds of property are not as closely tied to division of labour as Marx and Engels supposed. Different kinds can coexist with' much the same division of labour. Formal ownership is not the only – or always an effective – means by which control over niches within the social division of labour can be secured, and nor is it the only source of power. However, public ownership or nationalization in the guise of 'social ownership' is not the only alternative to minority, private-capitalist ownership, and vice versa. It is of considerable political, as well as theoretical, importance that we recognize the possibility of majority or universal private property and restricted collective ownership. Consequently privatizing former public assets need not take the form of handing them over to private capital, nor need abolishing private capitalist ownership mean nationalization; bringing control under non-capitalist, non-public organizations, such as cooperatives, is another option.[12] Externalities cannot all be either neatly internalized by appro-

---

[12] A recent survey by Peter Saunders showed that many people were against both

priate adjustment to private-property rights or abolished by the creation of social property. They are a reflection of the intractable complexities and interdependencies of physical, biological and social systems.

Any discussion of ownership and control is inevitably highly abstract; it simply does not cover enough aspects of subjects and organizations to allow us to make confident inferences about the kinds of concrete behaviours associated with each ownership form. This limitation has frequently been ignored with the result that the range of behaviours and outcomes possible within each form has generally been underestimated. Stereotyped contrasts between private and public activities, in which the former are associated with markets, revenue dependence, competition, profit and loss, and the latter with bureaucracy, tax-financing, planning, a lack of competition, and so on, radically underestimate the range of existing and possible behaviours. In practice, all these characteristics can be permuted in various ways. The standard public-versus-private contrast obscures the fact that state-owned firms can produce for markets. Privatization does not necessarily introduce competition. Decentralization and centralization of control can each produce similar problems whether the enterprises are capitalist firms, co-ops or public enterprises. Many of the assumed 'natural' characteristics of certain organizational forms are avoidable or can be modified by legislation. The effectiveness of organizations depends on management methods, how jobs are defined and coordinated, how information is handled, how learning and innovation are encouraged, how members are motivated, and so on. As Burawoy and Lukacs' comparative work on US and Hungarian firms shows, these mediating processes can confound standard expectations of economic behaviour based narrowly on ownership and funding (Burawoy and Lukacs, 1985). One cannot assume either that public ownership will safeguard the public's interest, or, on the other hand, that public organizations will inevitably be run for the benefit of those who operate them and at the expense of consumer welfare (Maynard and Williams, 1984; Sayer, 1992b).

I am not saying that anyone can do anything in any context – ownership still sets limits and incentives. But given this range of possibilities there are great dangers of misattributions of responsibility for specific effects, for example by blaming property relations for what is due to bad management. We need to ask what it is about a certain situation that gives rise to particular results (Sayer, 1992a, p. 92). If

---

nationalized industries and privatization! Although there is no evidence that those people had in mind the restricted collective ownership of market socialism as an alternative to both, it could be, thereby resolving the apparent contradiction.

provision by the state is poor, we need to find out whether it is a consequence of ownership, lack of competition, mismanagement, under-funding, etc. To evaluate economic organizations properly, we need a theoretical approach which is attentive to their concrete, multi-layered and multiply-determined character. How we behave with respect to a particular resource depends not only on ownership but on whether it's something we come into regular contact with in our daily space, on whether we value it or are indifferent towards it, on the material qualities of the object in question and the values and norms of the society. Some of these considerations – particularly the last – may actually be influenced by property relations, but they also have a degree of independence from them. We note this not to disqualify abstract theoretical claims about the tendencies of certain kinds of property, but to qualify their use in accounting for concrete behaviours. Ownership is clearly still of great economic and political importance.

7

Non-Capitalist Economic
Systems Compared

Having isolated or abstracted out different modes of coordination and
ownership and examined them in turn, we now briefly sketch some
major ways in which they combine in non-capitalist economic systems.
This draws together the themes of the previous chapters and helps us
assess whether particular conditions and problems experienced under
capitalism are unique to it; to understand what *is*, it helps to understand
how things could be otherwise. Outlines of major characteristics of petty
commodity production, market socialism and state socialism are
included for this prupose. The first two of these only exist in limited
forms alongside capitalist and state-socialist modes of production, but
they provide important 'contrast cases' because they use markets but
lack minority private property and hence class in the sense defined in
chapter 3. Although the analysis is at a more concrete level than the
previous sections, it is still quite abstract. Particularly in the cases of
petty commodity production and market socialism, I therefore stress
that these are ideal types, intended as models to be used for counterfactual purposes in assessing theories of capitalism; real economies are
messier and incorporate several systems (Hodgson, 1988).

I shall then compare these different types of economy on a number of
points: first, decentralization of control and resulting lines of conflict
and economic rivalry; secondly, the forms of economic surplus and
profits; thirdly, exploitation; fourthly, conflicts between the logics of
exchange value and use value; fifthly, economic autonomy and worker
security; sixthly, capacities of the systems for innovation; and seventhly,
problems of markets. In this way we provide an overview of the themes
of the last five chapters.

Petty Commodity Production

Petty commodity production (PCP) is particularly interesting with respect to the relationships between ownership and division of labour. We can define it as a system of commodity production by competing self-employed workers, or at least competing households. As such, in its pure form, it is classless; all workers are owners of their means of production.[1] There is private ownership, but on a universal basis rather than the minority basis of capitalism. While there are product markets, labour markets and capital markets are hardly developed at all. Profits can exist, though they are generally limited by the extent to which the self-employed can exploit themselves; petty producers do not have significant capital in the Marxist sense because they lack employees to generate surplus value.[2] Equally, they lack employees because the means of production are not concentrated into large enough masses. The social division of labour is more developed than under feudalism but the small scale of the units of production affords little or no scope for technical divisions of labour and little scope for technological development. The development of the forces of production is therefore 'fettered' by the social relations of production. In practice, there may indeed be some wage labourers, but many of these may be employed not by capitalists to produce commodities for sale, but as servant labour, servicing their employers' consumption.[3]

PCP is not a robust or stable form of organization that could have stood on its own and become as widespread and dominant as capitalism, for it has inherent weaknesses which make it liable to break down, unless highly regulated. Competition between petty producers is likely to produce winners and losers, with the losers becoming propertyless wage labourers. In some cases, they may be driven out of business altogether, but as we saw in chapter 5, petty producers with inferior means of production or relatively unremunerative lines of business may be better off opting to work for others as wage labourers. The winners can employ these wage labourers on condition that they produce more

---

[1] There may be patriarchal domination within the small production units, however.

[2] Petty commodity production therefore stands some chance of meeting the Lockean proviso for private property ownership, so massively transgressed by capitalism, that any person should leave 'enough and as good' for others.

[3] Historically, hundreds of thousands of young women who did not have the opportunity to inherit family property and had not married were proletarianized in this fashion. Theoretically, this implies: (a) that not all wage labour is capitalist wage labour; (b) that not all proletarianization leads directly to capitalism; and (c) that patriarchy can be a cause of proletarianization.

than they get in wages. The resulting profits can then be converted into capital or 'self-expanding value', and used to employ more workers. In this way, successful petty producers can become capitalists. In any case, even before this happens, exploitation can exist without class, as a result of inequalities in holdings of means of production, so exploitiation could be expected in a pure PCP system (Roemer, 1988). In addition, the limitation on the scale and cost of means of production placed by petty commodity production – its inherent 'dwarfism' – invites the formation of larger units by mergers and takeovers and the employment of wage labour. Marx recognized the force of these mechanisms: 'It is just as pious as it is stupid to wish that exchange value would not develop into capital, nor labour which produces exchange into wage labour' (1973, p. 249), and he was scathing about those who wished to preserve petty commodity production – perhaps on the grounds of saving the equality implied by universal private ownership – endorsing the claim of Pecquer that this 'would be ... to decree universal mediocrity' (1976, p. 928). The lack or limited development of capital markets[4] stunts the growth and transfer of resources to where they are most needed. Where these markets do develop they make a major contribution to the development of capitalism out of petty commodity production.

These voluntary or 'clean' routes to proletarianization and exploitation are of course not the only means by which capitalism became established; indeed historically, as Marx stressed, forcible expropriation seems to have been more important (1976, part 8).[5] However, Marx probably also downplayed other routes towards class formation because they don't cast capitalism in such an unfavourable light as does forcible expropriation.[6] Whatever their historical extent, the clean routes are important for understanding how exchange relationships between different parts of the division of labour can lead to the development of class relations.

Although PCP is vulnerable for these reasons, it can also survive in niches which capital finds too small or specialized to fill. Thus it persists in attenuated form within capitalist social formations, as some individual producers may have enough skills and means to production to earn

---

[4] In practice petty producers often borrow, though often to bridge consumption gaps rather than to expand. Indebtedness – often to other family members, and hence avoiding resort to capital *markets* – is a common feature of petty commodity production. My thanks to Wendy Olsen for discussions on this point.

[5] Recall that the relationships which follow from 'cleanly' originating exploitation are unlikely to remain clean (see chapter 5).

[6] For a more charitable view of Marx's analysis, see McNally (1993).

a living but not enough to employ others.[7] Even today there is a continuing competitive interaction and movement between PCP and capitalist organization, involving the movements of powerful skilled workers in capitalist firms leaving to go freelance, while other petty producers lose the ability to survive in competition with capitalist firms and become proletarianized.

Why dwell on this marginal form of economic organization? The reason is that it is helpful for thinking about markets separately from class, and hence for distinguishing capitalism from other market systems. Moreover, it helps to demystify the long tradition in liberal economics of blurring the distinction and attempting to pass off a model of petty commodity production as a model of capitalism – a practice which has obvious advantages for those wanting to present capitalism in the most favourable light (Schweikart, 1992). Writing in the 1770s, when industrial capital was in its infancy, Adam Smith could be excused for not distinguishing the two systems, although he certainly did not evade questions of class. However, the subsequent development of capitalism cannot be assessed on the model of the mutual benefits of exchange amongst bakers, brewers and butchers. The capitalism-as-petty-commodity-production model conceals capitalist social relations and the concentration of power they allow, overlooks the unprecedented development of technical divisions of labour, and generally helps distract attention from liberal economics' embarrassing silence on the nature of production and economic development, by reducing economics to exchange systems (see Schweikart, 1992, pp. 29–30). The case against the capitalism-as-PCP model amounts to far more than the point that enterprises differ in size from the corner shop to General Motors, and therefore have very different powers with respect to consumers; it is also that they have very different powers with respect to workers within the sphere of production. The labour contract is not a contract for the exchange of labour but for the selling of workers' labour power for specified periods, during which capitalists are obliged to get their workers to produce goods or services which have a greater value than the costs incurred, including wages. Furthermore, the neglect of production and innovation leads this kind of economics to miss the main sources of the dynamism of capitalism.

---

[7] There is much more to Roemer's analysis than this deliberately diluted account; for example, he also shows that given an initial unequal distribution of assets, some individuals will combine wage labouring with self-employment or self-provisioning. Capitalists are also differentiated into full capitalists, who do not have to work and whose income comes solely from profits arising from ownership of the means of production, and those who also have to work (Roemer, 1988).

Market Socialism

Consider further our ideal type of market socialism, introduced in chapter 3, in which workers own and control the enterprise in which they work, but not other enterprises, in the manner of restricted social ownership as defined above. All enterprises produce commodities for sale in competitive markets in which economic power is highly decentralized.[8] There are therefore competitive product markets, but no labour markets and hence no classes in the sense defined in chapter 3, since all workers or the vast majority would be owners, not wage labourers.[9] There would be no exploitation via class but it would still exist via unequal exchange between cooperatives with differing amounts of means of production. A substitute for the capital market would be needed to avoid making enterprises dependent on internally generated investment funds, to compensate for differences in capital intensity between enterprises, and to enable investment to respond quickly to new patterns of demand and to take place in new ventures (Brus and Laski, 1989, p. 80). Just how the provision of loans for investment is organized has important implications, for dependence on outside funds attenuates workers' control, though depending on who or what organization is providing the credit, this may not be a bad thing. In practice, some form of joint worker ownership and state-bank ownership (whether local or central state) may be necessary to overcome these problems. (Compare the models of market socialism suggested in Bardhan and Roemer, 1992; Roemer, 1992; Estrin and Le Grand, 1989. See also Horvat, 1982; Miller, 1989; Schweikart, 1987; Przeworski, 1991).

As we saw in chapter 3, workers under market socialism have strong control over their immediate work, but their economic survival is precarious since they are subject to competition and the vicissitudes of *ex post* market coordination. In short, the anarchy of the social division of labour remains. However, the upheavals in the development of the social division of labour are likely to be less severe than under capitalism. Abolition of the capital–labour relation simultaneously reduces the

---

[8] Former Yugoslavia was often described as having market socialism, but the differences are great enough to cast severe doubts on its being treated as an example of what we have in mind. Closer cases are far more modest in scale and to be found in cooperatives such as those of the Mondragon group in the Basque country (Horvat, 1982; Thomas and Logan, 1982).

[9] This kind of organization has sometimes been called 'syndicalism', but the term has also been used to cover other arrangements, many of them quite different from our ideal type, and carries too much unwanted historical baggage to be useful here.

enormous disparity in mobility – and consequently power – between investment and labour, which contributes so much to uneven development (Harvey, 1982). Capital may decide to close a plant and employ a new workforce on the other side of the world, but obviously this is not an option for the members of a co-op. Nor, quite clearly, can cooperators gain – as individual capitalists can – by cutting wages. Income differences within enterprises may be less than under capitalism, but remembering the levelling effects of markets, then unless there is an incomes policy at the social level, inter-enterprise differences are likely to be large because of the lack of a labour market (Brus and Laksi, 1989, p. 100). Labour would also need some protection from the lottery of the competitive markets, for there would inevitably be bankruptcies and some workers would not be able to join cooperatives. Market socialism may abolish class as we have defined it, but it has its own sources of inequality.

Our ideal type of market socialism cuts many corners, especially the particular form of ownership and control as regards workers, creditors and consumers, and the unavoidable need for central state influence. But it highlights the extent to which ownership of the means of production and the mode of micro-economic coordination of the division of labour can be separated. While the ideal type takes markets to be the only mode of coordination to be required, other, supporting forms of horizontal control – such as relational contracting or industry associations – are likely to be needed, as the Mondragon cooperatives show (Thomas and Logan, 1982). In any case there will still inevitably be activities which need to be centralized under the state. In practice, varying patterns of economies of scope and scale would encourage mixtures of centralization and decentralization. The material divisions and interdependencies of technical and social divisions of labour would afford additional scope for local differences in power, to the advantage of certain strategically placed groups (see Burawoy and Lukacs, 1985). Nevertheless, as a starting point it helps to abstract from these effects and consider a heuristic model of 'pure' market socialism so as to highlight the effects of major economic structures.[10]

State Socialism

While the market mode of coordination can be combined with more than one kind of property ownership, comprehensive coordination by

[10] For further discussion of the problems of cooperatives see Sayer and Walker, 1992, pp. 260–2; Estrin 1989; Schweikart, 1987; Elster, 1989; Miller, 1989; Thomas and Logan, 1982.

central planning can only be combined with public or state ownership. Having already discussed these we can be particularly brief.

Under state socialism, the economy is controlled according to the Plan, access to the means of control being the Party. As long as there is a complex division of labour and scarcity there will be differences of interests among producers and between producers and consumers. Although the planning authority does not own the surplus product as do capitalists, it controls its distribution, giving rise to a basic antagonism between the direct producers and those responsible for planning. In view of this, Szelenyi and many others have argued that this creates a 'basic class distinction . . . around the processes of production and expropriation and the allocation of surplus, around the relations of production (Szelenyi, 1988, p. 30).

Many Western socialist critics of state socialism have seen its deficiencies almost entirely in class terms, implying that the workers must liberate themselves from the power of the bureaucrat class, and take direct control as a free association of producers, abolishing the state. But while the class or quasi-class character of the bureaucracy is clear, the conclusions drawn from this are hopelessly awry; indeed, this is equivalent to the mistake of supposing that the only obstacles to workers exercising collective *ex ante* control over the economy are the state and private property in means of production. Democratic control by a 'free association' of workers is impossible, because 'centralism is the inevitable price which must always be paid for the abolition of the market without the prior eradication of its preconditions' (Selucky, 1979, p. 34). The quasi-class character of state socialism derives from the necessity of a vertical division of labour of control for centrally coordinating a complex horizontal social division. As Bahro has demonstrated with considerable force, 'actually-existing socialism' did not remove the division between the labour of strategic planning and control of a range of activities and the 'subaltern' labour of execution of particular tasks (Bahro, 1978). Alienation therefore worsens rather than eases in this situation. The token nature of public ownership, compared to that of property over which people actually have some power (e.g. private agricultural plots) invites free-rider problems and a lack of a sense of responsibility for public property (Aganbegyan, 1988). Work is no more meaningful when driven by plan targets than by profits and exchange-value criteria.

In a highly complex economy, replacing market coordination by a priori planning does not increase the transparency of the relationship between production and consumption. If anything the contrary, for a priori control of such an economy has to be centralized, which means that the plan is interposed between producers and users, direct feedback from the users through prices is cut out, and the producers become accountable to the central planners. Producers are monopolists operating in a sellers' market, with centrally determined prices, and central ('vertical') control of inter-enterprise relations obstructs their 'horizontal' relations, making low quality and poor coordination almost inevitable. In other words, production is neither for profit nor need but for the plan. Likewise production is valorized not by users through exchange but by the plan. Yet consumption is still atomized, as under capitalism and other market-coordinated systems, but follows a different logic from that which guides production: consumers attempt to spend their money according to their valuation of the use value of products, but they have little choice between alternatives, and where they can say yes or no with their money there is little direct incentive for producers to respond (Nove, 1983). Workers are alienated from their wages, since producers do not have to respond to spending patterns but to the plan (Selucky, 1979, p. 39). In this schizophrenic form of coordination, workers 'are left with the interest of maximizing their incomes, while bureaucrats function as a ruling class having an interest in accumulation' (ibid., p. 36). In the former Soviet Union, disequilibrium, the absence of competition and the continual failures encouraged the development of a major informal economy of barter, black markets and corruption. Ironically, by providing short-term solutions to problems, this informal economy propped up rather than undermined the planned economy. Cooperatives were also benefactors of these disequilibria, often making large profits denied to those in state enterprises. Consequently, like black marketeers in a wartime economy, this often made them objects of contempt rather than hope.

Crucial to the combination of worker security without autonomy was the policy of prohibiting unemployment by invariably rescuing enterprises from bankruptcy. Workers were therefore shielded from conflicts inherent to the uneven development of the social division of labour, though they also indirectly suffered the costs resulting from this fettering of the development of the forces of production. This fuelled inflation and allowed situations where hardly any enterprises showed losses at the same time as the national economy was in deep trouble (Kornai, 1986, p. 92).

Like capitalism, state socialism therefore had its inherent contradictions, and while the immediate cause of its downfall appears to have had more to do with the weakening of the political mode of coercion (Szelenyi and Szelenyi, 1994), the economic contradictions were perhaps the major conditioning element. Despite its remarkable and almost total disappearance, the structural problems of state socialism present lasting lessons for political economy.

The Systems Compared

Having outlined the main characteristics of these non-capitalist systems, we can now draw out some of the key similarities and contrasts with capitalism. (For fuller comparisons, see Buchanan, 1985; Kornai, 1986; and Putterman, 1990).

1 Decentralized control, producer–user conflicts and economic rivalry

Where there is an advanced division of labour or catallaxy, its control is necessarily somewhat decentralized and fragmented. Thus, even under centrally planned state socialism, the division of labour and knowledge means that significant degrees of power and autonomy remain with specialist producers in the various parts of the economy, and cannot be entirely subordinated by the centre. So fragmentation and dispersion of control are not restricted to capitalist economies or even to market economies, but exist under state socialism too. This reinforces our contention that the effects of the social division of labour are not to be reduced to those of one of its modes of coordination – markets – as many theorists assume. In any case, conflicts within market societies often have little to do with markets *per se*, and result purely from the fact that individuals can do many things without constraint or reference to others, even though they may often create problems for them (Lindblom, 1977). Such problems might be said to result from 'individualism', though this is more weakly and less exclusively associated with capitalism than often thought (Abercrombie et al., 1986).

Conflicts between producers, and between producers and consumers or users, are especially important. Producers don't necessarily know what is best for consumers. Such conflicts are by no means restricted to capitalism but would exist under market socialism and petty commodity production. Under state socialism, while enterprises did not compete in product markets for consumer spending, they competed in a more politicized way for their budgets from the centre. Whereas producer power over consumers is enhanced in state socialism by the monopoly organization of production, it is limited in capitalism and market socialism by hard budgets and competition (Kornai, 1986). Though the Left tends to be reluctant to admit it, competition can produce benefits in terms of lowered costs and improved use values, and (often but not necessarily) by favouring practices and organizations with superior efficiency and service.

2 Economic surplus and profits

In market socialism and state socialism, workers have to produce a
surplus over and above that which they consume, to cover investment
and transfer payments to those unable to do paid work. The systems
differ in terms of who controls production and distribution of the
surplus. Under capitalism it is obviously controlled by the owners of
capital who can appropriate it for luxury consumption and/or use it to
guide investment, regardless of the interests of the workers they employ.
Under state socialism the ruling elite is able both to control the
production and to appropriate some of it for itself. Market-socialist
enterprises need to produce a surplus if they are to meet competitive
pressures and invest, though they do not have to produce to support
a ruling class. In any of the systems, creditors, whether public or
private, can also gain control over the distribution and appropriation of
surplus.

Though strongly associated with capitalism, profits are not a uniquely
capitalist feature. At the level of whole economies they depend not on
the difference between purchases and sales but on the production of a
surplus. Profits are generally seen to have a motivating effect in encouraging suppliers to drive down costs and meet consumer demand. The
motivational qualities of profits depend not only on the extent to which
they reflect success in this respect but on producers actually receiving the
profits (or bearing the losses). If all profits were taxed away or all losses
written off by the government, the motivational properties would
disappear, and in turn the coordinating or regulatory mechanism of
profits would be lost. They therefore have two aspects, one concerning
their appropriation by particular groups, the other concerning their role
as an index of the relationship between supply and demand. Regarding
the former, the prime questions are who is able to appropriate profits,
and how or on what basis; regarding the latter, the function of profit as
an indicator of production performance and the satisfaction of needs
must be evaluated. The first question should not obscure the second, nor
should positive responses to the second question be allowed to suppress
the first. But this is usually what happens; on the Left the critique of
capitalist control and appropriation of profits is often taken to disqualify
arguments in favour of the qualities of profits (and losses) as indicators
of what is in short supply relative to demand, though of course demand
is not the same as need; on the Right, the regulatory qualities are
disingenuously used to conceal capital's class interest. As we saw in the
discussion of the invisible-hand thesis in chapter 5, the motivational
function may work better in non-capitalist market societies since in
capitalism the capital–labour relation limits the motivational effects of

markets on workers by denying them direct access to product markets and the earnings therein. A further feature of profits disliked by socialists is the fact that profit-takers can convert their profits into capital. But this is not automatic; worker-owners in market socialism could be regulated not to use their 'profits' to employ wage labourers and hence become capitalists.[11]

3 Exploitation

If we define exploitation in terms of situations in which one group or class can live partly or wholly off the labour of others, then it can be found in any market system, with or without class divisions, for it depends primarily on inequalities of holdings of means of production (Roemer, 1986). It applies not only to situations where propertyless workers are obliged to produce surplus value for capitalists as a condition of being able to work, but to workers (including the self-employed and cooperators) in enterprises with limited or dated means of production who have to trade with enterprises with superior means of production. In addition, in the case of unequal exchange in the 'narrow sense', low-wage workers whose products trade with high-wage workers (as in the North–South trade), are technically exploited (Emmanuel, 1972).[12] In principle, workers are free to leave any particular capitalist workplace and get a job in another, though as they lack means of production, they cannot generally avoid seeking employment somewhere. But labour markets are geographically localized and differentiated and highly segmented into submarkets, while labour mobility is limited by ties to localities. In practice, therefore, capitalists can take advantage of the difficulty workers often have in finding alternative employment. Hence the employment relation generally involves more domination than theories such as those of Roemer, which abstract from these ever-present 'frictions', concede. Employers can extract surplus value not only by virtue of possessing technology which enables workers to produce more than they consume, but by enforcing a discipline which makes them work harder than they would otherwise (Marglin, 1974). On the other hand, there are usually many opportunities and incentives for resisting domination and work intensification. Equally, this is not to

---

[11] Clearly, if cooperatives simply allocate income on the basis of percentages of their earnings after paying for overheads and working capital, then profits do not appear separately from wages.

[12] This does not mean to say that the exploitation could be ended by simply raising the wages of workers in poor countries, for their wages are limited by the low level of development forces of production: wage levels are constrained by the need to invest out of a limited revenue from limited output. See Larrain, 1989.

say that the absence of capitalist discipline might make cooperatives less productive; co-owners might actually work themselves harder.

4 Exchange value and use value

In all three economic systems, as in capitalism, money functions as a medium of exchange, representing in abstract form the value of products and services and the power that derives from ownership of them. With the development of credit and capital markets, possibilities for making returns without the intervention of production of use values expand, and in the circuit of financial capital, exchange value 'splits off' from the production and circulation of use value. Some capital markets may be regulated and conservative, with banks and other major investors having fairly stable patterns of borrowing from and lending to known, established clients, whose ownership is safe from challenge. Alternatively, at the other extreme, capital markets may be highly liberalized and speculative, with investment being responsive to short-term exchange-value indicators with few spatial or institutional constraints, and with company ownership being radically separated from management and firms being effectively permanently up for auction (Crouch and Marquand, 1993). Consequently, as capital markets develop, the process of uneven development shifts up a gear, allowing the accumulation, concentration and rapid switching of money capital between a vast range of different outlets, on an increasingly global scale. Thus, especially in more liberal regimes, capital appears to free itself even further from the material and parochial constraints of particular kinds of production, workforce and locality. However, as Harvey (1982) has emphasized, this attempted liberation can never be carried through completely; 'fictitious capital' always depends on the continued expansion of capitalist *production* somewhere. Although there are likely to be capital markets under market socialism, albeit highly regulated ones, the absence of the capital–labour relation significantly restricts the tendency of money capital to liberate itself from use-value production. While it may make sense for cooperatives to make investments outside their own enterprises, they are unlikely to do so in a way that puts their own work at risk. Thus, while there is still uneven development, it is likely to be a slower-moving process than under capitalism.

All the economic systems discussed exhibit conflicts between formal rationality and substantive rationality, and between exchange value and use value. Under all of the market systems, the goal of production becomes expressed in exchange-value terms in the form of profit, with use value becoming a means to the end. What might otherwise be substantive goals for providing certain use values become subordinated to the need to minimize costs and maximize revenue. On the other hand, ignoring formal rationality in meeting substantive goals might lower efficiency so much that the goals could not be met. Competitive pressures may in many cases lead inadvertently to use values being improved and provided to an increasing number of consumers, in accordance with invisible-hand explanations. Yet the dominance of exchange value as a regulator of economic activity also leads to deeply irrational behaviour, as in crises of over-production. It is therefore common to find unused labour power (unemployed bricklayers) and resources (surplus bricks) which cannot be deployed to satisfy unmet needs (houses for the homeless) because the exchange value conditions are not right. Under our ideal-typical market socialism, these problems are likely to occur on a smaller scale than under capitalism. This is due to the muted tendencies for money capital to be liberated from constant capital (or whatever one wants to call it) and hence to be able to fuel rapid investment and disinvestment, and due to the lack of need to pay for capitalist consumption. Under centrally planned state socialism, the problems are reversed in that there is a deficit of formal rationality, there being no competitively established prices, so that the centre tends to respond to political lobbying without having a means for assessing the relative scarcities and costs of different courses of action. Enterprises face weak incentives to economize, since they are protected from the threat of bankruptcy and because the planning process inadvertently encourages firms to over-order and hoard materials so as to be sure of meeting plan targets. As students of such economies have found, attempts to use non-price forms of valuation of output generate major irrationalities (Rutland, 1985; Nove, 1983; Lavoie, 1985).

5 Economic autonomy and worker security

Work security is greatest in state socialism because of the prohibition of unemployment. This comes at the price of limitations on individual liberty and mobility, and restrictions on economic restructuring, since workers have to be redeployed and reallocated rather than simply made redundant. Regarding workers under capitalism, most radical analyses emphasize their class positions as the source of their insecurity, while liberal analyses stress the effects of competition and shifts in demand, and 'friction' in the mobility of labour and capital. In fact, both sources are important in capitalism. Under PCP and market socialism, where workers are not propertyless, the second source of insecurity remains, showing that the two causes are separable. Marx's double-nature argument recognizes that individual workers' lack of control over their work is partly due not only to their class status as wage labourers, but to the increased scale of technical divisions of labour – a point also recognized by Weber. Marx nevertheless failed to relate workers' lack of autonomy to the growth of the social division of labour, and Marxists usually overlook the fact that these non-class sources of powerlessness are to be found in other kinds of advanced economy too.

6 *Innovation*

Competitive markets encourage experiments in economic behaviour and organization (Nelson and Winter, 1982). The inherent 'dwarfism' of PCP limits anything but small-scale innovation. Under capitalism and market socialism, innovations can be introduced without concern for any negative impact on existing external, 'third-party' activities and workforces, though market-socialist enterprises cannot be expected to prejudice their own workforces' interests by innovating, as capitalist enterprises often do. On the other hand, competitive, decentralized control can be inimical to the development of major long-term projects, and state organization may be required to support these. State socialism has an obvious advantage where such large-scale projects are needed, but the guarantee of full employment and the attempt to control economic development centrally inhibit 'bottom-up' innovation (Sayer and Walker, 1992, p. 259; Nove, 1989). In principle, socialist planners can take into account anticipated negative consequences for existing activities of planned innovation; while this happened in practice where employment was concerned, environmental externalities tended to be ignored.

7 *Problems of markets*

As Nove points out, the problems of macro-economic regulation, *laissez-faire* and monopoly are much the same under capitalism and market socialism, although market socialism would not have the desta-bilizing influence of unregulated (international) capital markets. Neither system can guarantee that the growth paths of the various parts of the economy will be harmonized. Growth ensembles may emerge by a mixture of accident and partial state design through the development of suitable modes of regulation (Aglietta, 1979; Lipietz, 1987; Jessop, 1990). In both capitalism and market socialism there is a need for state support and centralized provision of many services, which isolated producers, whether capitalist firms or cooperatives, need but cannot provide. For these reasons, real economic systems are inevitably less 'pure' than our ideal types acknowledge. Finally, the inegalitarian and egalitarian qualities of markets noted earlier exist somewhat independently of the social relations of production; for example, commodified health services have regressive effects regardless of whether the providers are capitalist firms, cooperatives, or the state. Only the state has the option of de-commodifying the services by financing them from taxes and providing them free at the point of delivery, though even then it can subcontract their provision to private firms, cooperatives or other organizations. However, while inegalitarian tendencies and uneven development are to be expected in any advanced economy, they are greatly amplified in capitalism by the class position of workers and the associated scope for exploitation, and by the hyper-mobility of money capital.

Conclusions

In the last five chapters we have presented a disaggregative analysis of modes of coordination, types of ownership and economic system. This provides material for conducting thought experiments and intersystemic comparisons which enable us to evaluate radical explanations and critiques of capitalist development. In particular, this kind of analysis distinguishes those features which are unique to capitalism, or occur in a particularly distinctive form of capitalism, from those which can occur under systems with different social relations of production and modes of coordination, thereby sharpening understanding of capitalism's specificity. It therefore helps us identify more precisely the causes of particular outcomes within capitalism; for example, whether a certain effect is a consequence of minority ownership of the means of production, or of the existence of markets or an emergent property of their combination, or whether it is inevitable in any advanced industrial economy because of its character as a catallaxy. Petty commodity production and market socialism may be miniscule alongside capitalism, but they provide useful comparisons. Capitalism is the most pervasive and successful form of commodity production, but it is not the only one. This is a matter of considerable political, as well as theoretical importance. A disaggregative analysis further helps us assess both the feasibility of the alternatives implied in critical standpoints, and their desirability, by indicating the problems that non-capitalist economic systems are likely to generate. By contrast, if we treat particular modes of coordination or types of ownership as covering other dimensions of variation too, as in the plan-versus-market and private-versus-public frameworks, then we are likely both to misattribute causal responsibility for particular effects and to underestimate possible alternatives.

The inevitable diversity and dispersion of knowledge and power in advanced economies have effects which are strongly mediated by particular forms of ownership and modes of coordination. These forms of organization react back upon economic activities, shaping them in particular ways. But as we saw in chapter 3, though adaptable, economic activities cannot be organized in just any way: food can't be grown overnight or stored indefinitely without being treated; railways can't be run anarchistically if they are to be safe; it is easy for the electorate to vote on capital punishment but not on how the pharmaceuticals industry is to be run, and so on. In theoretical discussions we may abstract from the enormously varied material properties of economic activities, but in moving to a more concrete level of analysis we would have to pick up the effects of these material differences in variations in economies of scale and scope, in differences between private and public goods, and find a finer set of implications for different forms of property and modes of coordination. Since organizational forms have to be responsive to what they are organizing it doesn't make much sense to try to make a case for 'market socialism', 'democratic socialism' or whatever as a universally applicable principle of organization. Not surprisingly, actual economies combine several different forms of organization.

Some readers may have found the highly qualified character of the evaluations of the last four chapters depressingly equivocal, an impression probably created by my tendency to balance criticisms with countercharges using 'on the one hand . . . on the other hand' constructions, instead of declaring unambiguously in favour of one system or mode and against all others. I certainly do not see this as a sign of vagueness or indecisiveness. Far from abstaining from judging economic systems, I have evaluated them in more detail than is often done, making more specific favourable and negative comments. By contrast it is general condemnations or unqualified advocacy which are too vague. Nothing is to be gained from pretending there need be no trade-offs. To their credit, advocates of market socialism tend to recognize that compromises have to be made, for example between efficiency and equity (Estrin and Le Grand, 1989). The emphasis on the intractability of the social division of labour in advanced economies may seem especially gloomy, in contrast to traditional hopes of collective *ex ante* control of the economy. However, the latter aim is hopeless rather than hopeful; freedom is not advanced by ignorance or dismissal of constraints but by their recognition. In fact, our disaggregative analysis shows that economic possibilities are less constrained than is commonly assumed on the Left. There are more alternatives than markets and planning. Moreover, many of the characteristics of organizational forms often taken as 'natural' are avoidable or can be modified by regulation, and we have consistently stressed that concrete forms and behaviour are partly dependent on features other than those theorized here, such as strategy.

8

The Critique Applied: Explanations of Uneven Development

In the preceding chapters I have addressed some major problems in radical political economy in the abstract. While some of the arguments might already be accepted, they are easy to forget when we confront concrete instances of capitalist development. I therefore now want to indicate how the critique applies to attempts to use radical political-economic theory for explaining more concrete issues. The concrete character of urban and regional development – arising as it does from the interaction of many processes normally dealt with separately (and hence out of context) – makes it a testing area of application for such theory, forcing one to consider the many complex interdependencies between different spheres of action. Furthermore, the salience of uneven development and inequalities in urban and regional processes makes them a particularly challenging subject for mainstream or liberal approaches, which generally evade such issues.

Considerable progress has been made in understanding these subjects; radical political-economic theory was not only applied to but developed to deal with hitherto largely neglected topics such as urbanization, regional development, industrial restructuring, the flow of capital in and out of the built environment, and housing crises (e.g. Allen and McDowell, 1989; Bluestone and Harrison, 1982; Harvey, 1985; Hudson, 1992; Lamarche, 1976; Lojkine, 1976; Massey and Catalano, 1976; Massey, 1984). In urban sociology Weberianism moved to the left and entered a constructive dialogue with Marxism (e.g. Saunders, 1983; Pahl, 1984), and even researchers politically unsympathetic to Marxism acknowledged its explanatory power. However, by the late 1980s the changes we have described in radical social science generally, such as the move into middle-range theory and the diversification of radical concerns away from capital and class, were felt in urban and regional studies too.

While these developments were hotly debated (see David Harvey's critique and the responses, 1987), few detailed their dissatisfactions with the abstract political-economic theory used in explaining uneven development; Stuart Corbridge's *Capitalist World Development* is one of the few exceptions (1986). At the same time, acknowledgements that the rise of neo-liberalism might pose a challenge have been remarkably few (e.g. Saunders and Williams, 1986; Saunders 1989).

In evaluating the more abstract political-economic theory it is thus necessary to go back to the older literature. Some of the key examples of the more explicitly Marxist literature which I shall cite, from authors such as Harvey, Scott and Lojkine, are therefore somewhat dated. However there has been a limited amount of recent theoretical work on the political economy of space, such as that of Sheppard and Barnes (1990), and Eisenschitz and Gough (1993) have published a Marxist-informed analysis of radical local economic policies which is particularly good for revealing critical standpoints.[1] The influence of the frameworks developed in the early work may have weakened but they are still apparent in recent middle-range literature such as the British locality studies (Cooke, 1989; Harloe et al., 1990). Though now neglected, the early theory is still of major importance and continues to warrant close attention. Since we cannot do without some kind of political-economic theory we have no alternative but to either accept it as still relevant or change it.

Our target here is a critical social science which has rarely examined its own critical standpoints – and indeed apparently sees no need for doing so – and which, not surprisingly, tends to be based on infeasible and often undesirable implicit alternatives. In short, the critical standpoints of radical urban and regional political economy embody an extreme form of constructivism. This is coupled with and encouraged by a systematic misattribution of causality, which we shall term 'class reductionism', in which processes largely attributable to division of labour and modes of coordination are attributed to capitalist social relations of production. I shall argue that many of the problems which the literature identifies are not entirely unique to capitalism but are particular forms of problem which are endemic in any conceivable advanced economy.

In using the term 'class reductionism' I am extending it beyond a usage which has become associated with post-Marxist and feminist critiques. Most critiques of the tendency in Marxism have focused on its neglect

---

[1] Harvey's much-cited *The Condition of Postmodernity* (1989) provides a recent illustration of the use of radical political-economic theory, this time in conjunction with cultural theory, though one which has been evaluated primarily in terms of the attempt to link the two rather than in terms of its economic theory.

or demotion of non-class sources of power and social divisions such as patriarchy, racism, nationalism, homophobia and so on (e.g. Row-botham et al., 1979; Giddens, 1981; 1985; Corbridge, 1986; Hartmann, 1979; Massey, 1991; Laclau and Mouffe, 1985; Walby, 1986). I am very much in agreement with these critiques; they expose differences, inequalities and forms of domination, and indeed whole areas of experience about which Marxism – and indeed mainstream social science – had little to say. Nothing which follows contradicts these post-Marxist contributions. However, I think they too can benefit from a consideration of the further kind of class reductionism that ignores the materiality of divisions of labour.[2]

Related to class reductionism is an issue which was of great importance to radical political economy in its critiques of mainstream theory – historical specificity. Mainstream social science and economics in particular were taken to task for treating processes and practical logics which were specific to capitalism as if they were universal and transhistorical; there were not simply economies, cities and states but capitalist economies, cities and states, with ways of acting which were peculiar to them and had to be understood as such. I shall argue that it is doubtful whether radical political economy succeeded in establishing what was historically specific about them, because of a tendency to assume that everything that existed within capitalist social formations was uniquely capitalist, instead of leaving this as an open question.

'Capital versus Community': The Privilege of the Producer

A common form of class reductionism is evident in the counterposition of community and capital, such as 'Liverpool versus the multinationals' (e.g. Bluestone and Harrison, 1982).[3] In this framework, the interests of communities or localities are seen as corresponding to their workers' interests and as counterposed to the interests of capital. The mobility of international capital and the deindustrialization of the last two decades, coupled with the dependence of localities on their economic 'base', gives this framework considerable force.

Certainly part of the reason for the insecurity of employment is workers' class position, their lack of ownership of the means of production, but it also reflects the vulnerability of any group to changes

---

[2] For example, Laclau and Mouffe (1985, p. 178) write as if the social division of labour did not exist.
[3] For fuller discussions of examples of class reductionism, see Morgan and Sayer, 1988, ch. 14, and Sayer and Walker, 1992, ch. 6.

in product markets. The one-sidedness of the capital versus community approach can be exposed by comparing it with an equally one-sided liberal view. From Adam Smith, through von Mises (1935) to the present day, some liberals have argued that the real boss of the worker is the consumer, while the owner or manager of the firm is merely an intermediary between them. From a Marxist point of view, this is an outrageous claim because it can clearly be used to distract attention from, or legitimize, the domination of workers by capital. But the two perspectives are not entirely mutually exclusive; we can fully accept that the capital–labour relation is one of domination while simultaneously recognizing that there is something in the liberal claim that the fate of the workers is also dependent on consumers, be they individuals or firms or other organizations. The fate of localities therefore also depends on producer–consumer relationships, and these cannot be cancelled out so we are just left with capital versus labour in the equation.

To be sure, it is capital that pulls the plug through disinvesting, and Bluestone and Harrison rightly attack the speed with which such decisions are often taken, leaving no time for devising alternative strategies. But this does not mean that capital is at the end of the causal chain; it may be responding to conditions beyond its control. Restructuring is often carried out for reasons which have little or nothing to do with labour, although it obviously impacts on it: labour is not necessarily a problem for capital.[4] Capital itself – no matter how big – is vulnerable to shifts in demand and technology (Morgan and Sayer, 1988), and can be hit by largely unforeseeable shifts. Workers do sometimes lose their jobs because consumption patterns change, and this is not restricted to capitalism but would happen in any market system. In the town where I live, thousands of workers were once employed making linoleum; among the reasons for the collapse of this employment was that consumers switched to other forms of floor covering (Bagguley et al., 1990, p. 42). Had the enterprises making the linoleum been worker-owned, they would have still been vulnerable to this shift. The insecurity of workers derives from the product market as well as the social relations of production.

In addition, the workers are likely to have interests which differ from those (usually unknown) local or distant people and organizations who consume what they produce. Thus, given the difficulty of finding alternative work, steel workers usually want to continue producing steel, while some consumers may want to switch to other kinds of material

---

[4] The assumption that labour is always capital's main problem was one of the chief deficiencies of interpretations of management and the labour process influenced by Braverman (1974).

which are substitutes for steel. Marxism's focus on people as producers leads it to dismiss the difference and support the producer. More generally, retaining high-cost producers when low-cost alternatives are available means that other kinds of economic activity have to be forgone which might otherwise be affordable. By contrast, liberals – with their preoccupation with people's role as consumers, and their tendency to focus on the coordination of the division of labour through markets at the expense of a concern with class – repeatedly undervalue the producers' interests and vulnerability, not least because they also frequently overlook their geographical localization and distance from alternative work and the dependence of communities upon them. But equally, precisely because there is a profound division of labour, we should not assume that a specialist group of producers' interests are in accord with the general social interest, or even with local interests. A recent brochure from the Centre for Local Economic Studies (1992) states that 'economic decisions should be taken as close as possible to the communities they affect'. Clearly this simply ignores the dispersion and interdependence of interests across the division of labour. 'Communities' here refers to the places in which particular producers live, as if no one other than these producers were affected or had any interest (e.g. the consumers of their products), and as if the people in the places in question have no interests themselves as consumers. The conflict people face between their interests in consuming the cheapest or best-value goods and their interests as workers producing commodities wanting to maximize their income is not wholly reducible to a class, capital-versus-labour dimension: it would exist under the kind of classless market-socialist society outlined above.

Here a further comparison with a former state-socialist society is instructive. In northern Siberia the Vorkuta coal mines have long been a seat of political opposition and power. Its workers were a force to be reckoned with by the Soviet government, pressing for reforms and later supporting Yeltsin. The Vorkuta miners' relationship with the state can be seen in class terms, as a struggle over the survival of the mines – which are relatively high-cost operations – against those who controlled the economic surplus. But there are also fundamental conflicts of interest across the division of labour which applied as much under communism as under the emerging market system – conflicts between these high-cost producers on the one hand, and energy users and the rest of the economy on the other. Coal users (both industrial and domestic) would have been better off having cheaper coal, and again, as in the capitalist example, the whole population could be said to have had to forgo possible benefits in order to subsidize the Vorkuta mines. Market regulation makes these conflicts of interest between workers in high-cost operations and other workers and consumers more transparent, and tends to resolve them in favour of the latter rather than the former, but it does not necessarily create them. In other words, markets may be divisive, but the divisions were already there under central planning by virtue of the division of labour and the pattern of opportunity costs.

Another twist to the neglect of the way in which divisions of labour really do divide labour concerns policy regarding restructuring in capitalist countries. In the late eighties it was common on the Left in Britain to counterpose 'restructuring for labour' to 'restructuring for capital' (e.g. Gough, 1986). This neatly sums up the inadequacy of class reductionism and its misrepresentation of the structures of interests in an advanced economy, for as we have seen, restructuring is not purely for capital, and it is not simply class which prevents labour from controlling its destiny. The slogan 'restructuring for labour' merely invites ripostes such as 'restructuring for which workers, for which part of the social division of labour?', or 'what about the interests of those who use the products?' (Rustin, 1986).[5]

This class reductionism or 'privilege of the producer' is also evident in the locality studies of the 1980s – not surprisingly, given that the localities were basically defined in relation to local labour markets (Cooke, 1989).[6] While localities depend heavily on their local economic 'base', this does not mean that local interests can simply be defined in these terms and treated as unitary. As Urry (1990) notes, the interests of people within a locality vary enormously in strength and kind, and it cannot be assumed that local attachments come first, so that peoples' interests can be represented territorially. A glance at a local yellow-page directory gives some idea of the multiplicity and diversity of activities and interests within localities, just on the producer side alone. Furthermore, as was eventually acknowledged in radical urban political economy, many urban conflicts have little to do with production relations, and more to do with racism, struggles over access to amenities and exclusion of unwanted activities and groups, and home-owner struggles to maintain and improve property values (Davis, 1991).[7]

---

[5] See Sayer and Walker, 1992, pp. 264–8, for further (sympathetic!) critiques of socialist strategies under capitalism.

[6] Some of the studies tried to counter this bias – for example, by focusing on gender divisions or housing and local-service issues – but the problems of the producer bias and the broader consumer interests were never adequately addressed.

[7] Castells' own work illustrates this shift, moving from a position of radical wishful thinking in which urban social movements were seen as 'forming cross class alliances around consumption issues', thereby helping 'to unify anti-capitalist interests' (Castells, 1978, p. 36), to the more open position of *The City and the Grassroots* (1983), with its studies of non-class groups such as gays, and where such attempts at forcing conflicts into a class-struggle mould were less in evidence.

Uneven Development and the Logic of Exchange Value

A persistent theme in Harvey's analyses of capitalist uneven development is money itself, which he discusses in overwhelmingly negative terms (Harvey, 1982, 1985). Money and markets bring with them the levelling effects of competition, pressure to develop formal, rational calculation, the associated conflicts between formal and substantive rationality, and the rise of exchange value as the measure of things and the regulator of activity. These pressures cannot be escaped as long as scarcity persists. This is not to suppose that scarcity is pre-social, for it is always relative to demands made. Capitalism's growth imperative ensures the maintenance of scarcity, but so did the drive to catch up to the West in state-socialist countries. The resulting need for economizing, for making trade-offs between substantive goals, and for formal, rational calculation, are increased under capitalism but are not unique to it. Competition and struggle are partly a result of scarcity, and are not reducible to effects of private ownership. There are always competing claims on resources and they would still be present under central planning or any form of democratic control (to the extent that the latter is possible), though conducted less in exchange-value terms. The way in which this happens under capitalism is distinctive and problematic, but as the Vorkuta example showed, it is likely to be difficult in other systems as well. For all the tensions, irrationalities and contradictions engendered by exchange value, it is hard to dispense with it in advanced economies without producing greater irrationality.

Once markets exist, it is difficult for economic activities to resist their logic. Thus special banks might be set up with use-value goals paramount (for example, funding particular areas or developments), but if they have to compete for their funds rather than getting them from government grants, they have to offer competitive rates of interest to lenders, which means charging market rates to borrowers. This pressure exists irrespective of the ownership of the bank, although different forms of ownership may allow some different interpretations of the constraints. It is not restricted to profit making or privately owned activities. As long as institutions' budgets are hard and they have to buy some of their inputs, they have to submit to the logic of competition, and to the pressure to evaluate their activities in exchange-value and not merely use-value terms.[8] They may still be able to meet some use-value targets that do not maximize or conserve exchange value, and may indeed be regulated by

---

[8] Harvey notes that the same dilemma is faced by landowners in deciding how to regard their property (Harvey, 1982, p. 421).

the state to do so, but such requirements are likely to put them under considerable stress. Moreover, the more pressing the exchange-value constraints, the greater the pressure to limit their service to those who can pay the most for it. Nor is it wholly irrational, for as we have seen (chapter 5), prices give a basis for comparing costs at a social level. It is nevertheless absurd to make blanket condemnations of exchange value and hard budgets since they encourage parsimony and greatly facilitate cost comparisons. Subsidies can only be selective and in the economy as a whole, consumption, including productive consumption, must be limited by available resources. Although credit may allow consumption to be extended beyond this, it is predicated on future growth.

The regulation of economies in exchange-value terms is exemplified in the role of profits, interest and rent. Like profits,[9] interest and rent have two sides, a distributional aspect concerning who appropriates them, and an allocational aspect, concerning how they regulate flows of capital and credit or land and buildings among competing bidders, according to relative scarcity. As with profits, radical political economy exposes and counters the tendency of neoclassical economics to treat the allocational side as a cover for the relations of appropriation on the distributional side. Interest and rent are not payments to 'scarce factors of production' but payments made within social relations in which the owners of money and land appropriate some of the value which workers produce over and above their wages. However, radical political economy tends to invert the problem by largely ignoring the allocational side, giving the impression that the rationing effects of interest and rent are unimportant, as if acknowledging them would necessarily mean legitimizing the current pattern of ownership of surplus money and land. But interest and rent can be taken by groups other than private capitalists, landlords or individuals, such as the state or an elected body, and this is likely to be better than abolishing interest and rent, for it allows the allocational benefits to be retained.

The rate of interest is an important regulator of investment. Thus, for example, Harvey writes: 'Resources are taken from the earth and land taken up at the periphery . . . at rates dictated by the prevailing rate of interest rather than in accord with some other conception of current or future well-being' (Harvey, 1982, p. 397). (Harvey says nothing further about what this other conception of well-being might be.) Credit could be controlled centrally so that the state became the main interest-taker, but zero rates of interest would hardly encourage efficient use of resources. One can imagine some credit or grants being given purely on use-value grounds, but this would have to be constrained by the scarcity

---

[9] See above, chapter 7.

of resources. Credit has to be rationed if inflation is not to be generated and the simplest and sometimes the best way of assessing the competing needs of thousands of seekers of credit is to make them bid for it, and charging interest provides a simple, abstract way of doing this. This is not a uniquely capitalist phenomenon. In a market-socialist society the thousands of competing demands for credit would have to be assessed, and while the substantive rationality of the proposed uses of the credit might be given more consideration, the problem of formal rationality and scarcity could not be ignored.

Similarly with rent: theorists such as Harvey, Walker and Murray demonstrated the class coordinates of rent in capitalist society, as against the apologetics of neoclassical treatments (Harvey, 1974; Walker, 1975; Murray, 1977). As with interest, we might question who is entitled to own buildings or land and charge rent for their use. This is especially so with rent, for the market rent for a plot of land is affected by its proximity to other developments taking place, with the result that its owner can benefit from windfall increases in the 'floating value' surrounding the plot without lifting a finger. The case for passing increases in rent revenues reflecting 'development gain' back to the wider community instead of being pocketed by fortunate owners is therefore strong. Major inequalities in income also grossly distort the effectiveness of the allocational process, allowing the rich seeker of a second home to outbid the homeless. A strong case can therefore be made for allocating land or buildings at subsidized rates to institutions (e.g. schools) or individuals whose needs are not reflected in their ability to pay. Were it not for these inequalities, rent would ration resources among users giving due weight to the strength of their preferences. Again, this distortion is not peculiar to capitalism, but is common to any money economy. For this reason, quite apart from who owns property and land, outcomes in urban land markets may be a highly imperfect response to needs and preferences. Nevertheless, the solution may not be to abolish land markets and rent but to equalize incomes, for a general policy of zero rents or rents unrelated to scarcity would lead to serious misallocations of resources. Thus among the many irrationalities discovered by Szelenyi in housing allocation in state-socialist cities using a combination of bureaucratic procedures and subsidized rents, was a grossly inefficient use of scarce central urban space (Szelenyi, 1983).

Finally, whether fair or unfair, exchange-value allocational mechanisms are cheap to operate, for all the complex and contentious issues of comparing diverse needs in terms of substantive rationality on the demand side, and relative costs on the supply side, are bypassed by the fact that everything is translated into the abstract measure of money values. A limited number of needs and demands might be assessed without markets, according to some set of substantive criteria, but this is costly and clumsy, and is likely to lead to disputes regarding how substantive merit and needs should be ranked, whereas markets allow such valuations to vary individually.

Harvey places great emphasis on the fact that in order to produce surplus value, capital must be fixed for substantial periods of time in particular places, even though economic circumstances may change and render them sub-optimal from the point of view of maximizing profit. Capitalists therefore have to decide between persisting with current commitments and making what are probably sub-optimal profits, and taking the risk of devalorizing existing capital in the hope that a new round of investment – usually in different locations – will yield higher profits. The pattern, pace and character of urban development depends on how these critical decisions are made (Harvey, 1975, 1982, 1985). While this is a real enough dilemma for capital, it is also a capitalist variant of a general problem of any society in which there is innovation and change. The dilemma is more acute under capitalism, where the very survival of individual capitalist enterprises depends on making the right choices. Socialist planners would not have to follow profit signals in making judgements about specific disinvestments and investments, but they would still be confronted by a dilemma of whether to continue with existing commitments or switch to new investments, and they would soon be in trouble if they ignored it (Pahl, 1977).

Another commonly noted expression of the problem of existing commitments is in terms of the increasing tendency for 'living labour' to be dominated by 'dead labour' under capitalism. This concerns: (1) the way in which accumulation under capitalism rapidly builds up an enormously increased amount of means of production, infrastructure – the product of earlier labour – relative to living labour at any time; (2) the way in which this enhances capital's dominance, both by increasing the volume of capital and the rate of surplus value, and by subordinating labour to the discipline of the rhythms and requirements of technology. Under state socialism the domination of living labour by dead labour continues in terms of (1) and to a certain extent (2), with the state replacing the role of capital. Under market socialism, workers escape (2) and can control and appropriate more of the gains from the disproportion in (1). It is interesting to relate this criticism of capitalism to that of the mobility of money capital, for it becomes clear that with less mobility of capital, and hence fewer new ventures, dead labour would dominate living labour even more, for *in situ* growth would be far more common.

The temporal trade-offs between developing now or waiting are often associated with others:

> The tension between fixity and motion in the circulation of capital, between concentration and dispersal, between local commitment and global concerns, put immense strains upon the organisational capacities of capitalism.
>
> *(Harvey, 1982, p. 422)*

But again these are capitalist variants of more general problems which put strains upon any advanced economy. Socialist economies also have to face equivalent problems: taking advantage of the existing, most productive sites and activities so as to generate most wealth may increase inequalities further, while diffusing the investment to backward sectors and areas to gain more equality may jeopardize overall growth. Opportunity costs and dilemmas of development remain (Toye, 1989). While decisions involving equity-efficiency trade-offs in centrally planned economies are likely to be more politicized, the lack of market prices make it hard to assess the costs.

Stronger versions of such dilemmas also occur:

> If a powerful bank holds the mortgage debt on much of the infrastructural investment within a territory, then it undermines the quality of its own debt if it syphons off all surplus money capital and sends it wherever the rate of profit is highest.
>
> *(Harvey, 1982, p. 421)*[10]

But a state bank in a socialist economy would also be torn between supporting development in an area or sector in which it had already made loans and commitments, and lending for other, perhaps more pressing needs elsewhere. Allowing soft budgets or postponement of loan repayments or just printing money in the hope of meeting all needs simultaneously would not remove material shortages and the trade-offs that would accompany them. The capitalist reliance on rates of profit and interest as a guide to investment and disinvestment produces many irrationalities, especially in more liberal regimes, not least because they are a poor indicator of future investment success. But the useful question is, how does it compare with alternatives?

Note also that there is some good in these dilemmas, for where there is a complex and spatially extensive division of labour, it is not in the general interest for investors – be they private or public – to ignore

---

[10] A more glaring example of contradictions between investments and current commitments involved manufacturing and public-service workers in inner London in the 1970s, who lost their jobs partly because – unbeknown to them – their own pension funds were being invested in property development which was displacing the traditional activities of the area (Counter Information Services, 1973).

alternative investment possibilities in other sectors or places to those in which they are currently involved. It is not sensible for an enterprise to continue investing in the same activities in the same places if demand for their products has fallen. Again, one-sidedly supporting producer interests can easily become a kind of conservative parochialism.

Urbanization

> Cities are, first of all, seats of the highest economic division of labour. In measure of its expansion, the city offers more and more the decisive conditions of the division of labour.
> *(Simmel,* The Metropolis and Mental Life, *in Wolff, 1950, p. 420)*

Capitalist urbanization presents a daunting challenge to any theorist. Conventional political-economic theories can abstract from space and isolate out particular relations or processes, be they those of class, circuits of capital or whatever, and elucidate their structure. But a theory of capitalist urbanization has to explain its extraordinary combination of order and disorder, its combination of dense networks of connections between related activities and intermixing of individuals, groups and activities having no relationship to one another. It has to theorize not only conflicts built into relations of dominance but the multitude of conflicts that occur between contingently related but co-located activities. In addition, any theory of capitalist urbanization has to confront its historical specificity – a matter of great significance in the new urban political economy's critique of traditional urban sociology. However, the new radical approaches only partly succeeded in identifying this specificity. Since urbanization is not limited to capitalism we have to distinguish those features of capitalist cities which are unique to capitalism from those which are not. Like the governance of technical and social divisions of labour in capitalism, the capitalist city has a 'double nature' which class-reductionist approaches miss.

One of the most important themes in urban political economy concerned the economic role of the state in providing 'general conditions of production' and 'collective means of consumption' (Lojkine, 1976; Harvey, 1975; Castells, 1978; O'Connor, 1973; Scott, 1980). These state-provided services, products and infrastructure are needed by capital but cannot usually be provided by capital itself. State spending may support private capital directly, for example by providing roads, or indirectly as in the case of education and training. State support for collective consumption, such as socialized health care or parks, benefits labour but can also indirectly support capital. The lessons here have

implications for state theory in general. Much was made of the contradictions involved in the relationship between state and capital in the
urban area. For example, while state spending might benefit capital, it
was ultimately funded by taxes on capitalist accumulation, so that
excessive demands might threaten the source of funding. Since private
capital dominates the economy the state has ultimately to maintain
capitalist profitability if it wants economic growth; the productive
potential of the economy can't be used without rewarding those who
control the means of production (Przeworski, 1991). Furthermore, it has
not only to maintain this profitability but avoid creating a crisis of
confidence, for example by making too many compulsory purchases for
urban redevelopment. The interests of capital therefore restrain as well
as require state action, and capital can put pressure on the state to
comply with its wishes.

These ideas were a major advance over mainstream approaches to the
state which took it to be benign and which analysed it in abstraction
from the social relations of production and other power relations.
However, the mechanisms identified are not entirely unique to capitalism. Worker-owned enterprises, just like capitalist firms, would have
difficulty providing public goods like infrastructure. Under market
socialism, in so far as worker-owners control most economic activity,
and in so far as the enterprises would have to remain competitive, the
state would have to support them, while ultimately being dependent on
them for providing the economic development to fund the necessary
taxation. This would give workers leverage over other interests and
produce a bias towards labourist policies. Thanks to the impact of
feminist and other critiques of Marxism, such a form of domination by
labour is no longer assumed to be benign.

Marxist theories of the state and urban development insisted that the
state in question was a *capitalist* state, and that its effects had to be
explained in those terms. Unfortunately, this tended simply to invert the
problem of ahistorical state theories. The problems of state action in
capitalist society do not derive only from the fact that it is a *capitalist*
state – or indeed a patriarchal or racist state: they also derive from the
fact that it is a *state*. Whether capitalist or not, states inevitably have to
regulate, discipline, tax and undertake surveillance of individuals, with
good and bad effects. These functions can take contingent capitalist,
socialist or patriarchal or racist forms, but any state has to do these
things. Not all the effects of a capitalist or patriarchal state are purely a
function of its capitalist or patriarchal form. Moreover, it is only if we
recognize this double nature that we can understand the way in which
neo-liberal attacks on the state met with some popular support.

No state in an advanced economy can avoid having a largely bureaucratic character, and of course as the Weberian literature shows, that has good consequences as well as bad, in so far as bureaucracies rationalize tasks, and act consistently and 'without regard for persons'. Democratic governments necessarily require bureaucracies to discharge their special-ized roles, catering for vast numbers of people and activities and collecting data on them. As with any organization, state bodies become power bases in their own right. Any critique of the state would therefore have to decide how far its merits and defects derived from its capitalist or patriarchal forms and how far from its bureaucratic and technocratic forms. In the eighties there was a shift of emphasis within radical critiques of the state from a concern with its capitalist character to a focus on the 'Fordist' character of state organizations (e.g. Murray, 1986): the latter could be interpreted as (unknowingly?) engaging with the bureaucratic side of the state more than its capitalist side.[11]

The way in which capitalist interests could block the development of state welfare spending and hold the country to ransom has been a common theme in radical political economy (Przeworski, 1985), and the *critique* of capital's power over the state tended to imply that in the absence of a capitalistic form or shell, the state would become both all-powerful and benign. This popularity of this critical standpoint on the Left waned markedly in the eighties as a result of the rise of a curiously disparate set of sources – neo-liberal critiques of the state, populist anti-statism (especially in Britain), Foucauldian analyses of discipline and surveillance, and Weberian theory of bureaucracy. With the aid of these theories it became clear that attempts to find bureaucratic solutions to social problems created new power bases with interests of their own; as Habermas puts it, 'The legal–administrative means of translating social-welfare programs into action are not some passive, as it were, property-less medium' (1987, p. 155).

Our earlier discussion of central planning and state socialism suggest that the state cannot become all-powerful without dire consequences. The difficulties experienced by state-socialist countries in controlling diverse activities remind us that the role of the state is limited by the division of knowledge in society. While states are well-suited to the running of natural monopolies like water provision, they are ill-suited to activities with limited economies of scale and which respond to varied and localized users. These are more efficiently organized on a decentral-

---

[11] Curiously, as far as I'm aware, this shift took place without any major reflexive critiques of the earlier analysis of the capitalist state. My guess is that the Marxist provenance of the concept of 'Fordism' might have helped to create the illusion of continuity rather than rupture, while the traditional Marxist aversion to Weber might have prevented radicals noticing the similarities of their critique of Fordism to a negative reading of Weber's analysis of bureaucracy.

ized basis, though not necessarily under capitalist control. The dangers of the state displacing popular, individual, local or communal control are now also taken more seriously on the Left. Such points suggest that Weberians such as Keane (1984, p. 4) are right in arguing for a less negative evaluation of the restraining effect of private interests on state power, though these need not necessarily be capitalist interests. While we should indeed worry about capitalist domination of the state, we should also worry about the dangers of an unfettered state.

But there are other constraints on the enlargement of state activity besides those of class power, and the role of the state cannot be understood entirely from an accumulation and class perspective. The state also coordinates parts of the division of labour and allocates resources among activities. How it does this is not purely a function of class. While there are activities which it can organize more successfully than other agencies, we should not automatically assume, in constructivist fashion, that this means it could successfully do a great deal more. Being both required by capital and resisted by it might count as a contradiction, but being able to do some things efficiently and others not is hardly one. Of course there are contradictions in capitalist urban development. Some might be removable. But not all contradictions of urban development under capitalism are unique to capitalism; some would be present in any kind of imaginable advanced society.

As with work on industry and uneven development, urban research has found the Marxist distinction between the technical and social divisions of labour enormously illuminating (e.g. Harvey, 1985; Lojkine, 1976). We can best identify something of the double nature of capitalist urbanization by relating this distinction to the time-geographies of urban processes. On the one hand, within firms, the technical division of labour is rationally and despotically controlled, achieving solutions to time–space coupling problems in the movement of people and materials to a degree unequalled throughout society. On the other hand, outside the factory or office door, in the social divison of labour, there are mainly *ex post*, individualistic and piecemeal solutions to the rationalization of movement in time–space, hindered by the monopoly properties of space, the fixity of the built environment, the fragmentation of land ownership and the fact that people's ability to make economical use of space and time is influenced by power relations. The contrast between the relentless rhythm of the shopfloor and the anarchic waste of the traffic jam speaks volumes. Both wage workers and non-wage workers generally have to organize the spacing and timing of their activities to fit in with capital's priorities.

In part the space–time patterning of activities in the city can be seen in terms of a contrast between ownership forms, the contradictions between private capital and the public sphere, between capital's time and space and labour's time and space. But from another viewpoint it can be seen in terms of divisions of labour, patterns of (dis)economies of scope and the problems of organizing them in a dynamic context. The reason why the organization of different parts of an office or factory is far more rational than that between it and other activities in the city is not just a consequence of class relations and the imperatives of capital accumulation, but of the differences in economies of scope in the two situations, or, in Hayekian terms, because the former is an economy* and the latter a catallaxy. It relates not only to the different organizational and social forms present (our dimension A – see above p. 63) but to the character of the activities being organized (B). The production of things like toothpaste or telephone bills lends itself to a greater rationalization of their time–space patterning than would the joint production of chalk and cheese or the daily life of the members of a household. Each capitalist enterprise tends to involve itself not in bundles of diverse activities with few possibilities for time–space rationalization, but in specialized, compatible activities where all the constituent elements can be focused single-mindedly on the business of capital accumulation. Even though the form of the division of labour in a non-capitalist system might change so that some activities currently separated might be combined and vice versa, their compatibilities and incompatibilities would make some forms distinctly more attractive on efficiency grounds than others. There would still be a chaotic social spatial division of labour and the interactions between the activities would still be dense and difficult to regulate.[12] In other words, we again have to steer a course between a voluntarism in which the division of labour can be moulded to any form without cost, and a determinism which treats the existing forms of division as the only possible ones.

Equivalent arguments could be made concerning links between other power relations and time-geographies of urban space. Women's ability to control the spacing and timing of their activities is limited by men in a variety of ways, through monopolization of the means of transport, through violence and through restrictions on women's roles inside and outside the home (Little et al., 1988; Mackenzie, 1989; McDowell, 1983; Massey, 1991; Valentine, 1989). Further divisions can be demonstrated for race and sexuality (Castells, 1983; Smith, 1990).

---

[12] The forces governing spatial interactions across urban spatial divisions of labour have some common features in all advanced economies. Simmel commented:

> The technique of metropolitan life is unimaginable without the most punctual integration of all activities and mutual relations into a stable and impersonal time schedule . . . Punctuality, calculability, exactness are forced upon life by the complexity and extension of metropolitan existence and are not only connected with its money economy (quoted in Kumar, 1978, p. 71)

The division of labour within a household and the spacing and timing of its members' activities reveal a great deal about power relations, but they also reveal something about the activities being coordinated. Even if you have a dual-career household in which the jobs are of equal status and pay and the domestic work is equally shared, differences in the space–time regimes of the jobs can cause space–time coupling problems. The difficulties that members of households face in juggling activities are certainly unequally shared because of gender relations, but there would still be some problems in the absence of gender differences because of the mundane fact that even a gender-neutral divison of labour would involve activities which are often difficult to coordinate in space and time. Similarly, while urban political and economic processes are generally dominated by white, middle-class men, the barriers to democratization and dispersion of power go not only beyond class to gender and race, but beyond all of these to technocracy and division of labour.

I have no intention of abandoning the idea that industrial and urban organization are profoundly influenced by the mode of production. Such features as the mobility of firms and the enormous financial power of property capital derive from specifically capitalist characteristics. But the need for dense interactions would persist in any developed socialist city. The similarities in urban form between capitalist and state-socialist cities attest to what they have in common – industrialism and an advanced division of labour. Nor, as some readers might suspect, does this kind of argument simply return to the early kinds of time geography in which the time–space paths of individuals were analysed largely in abstraction from their social relations and roles and hence reduced to a purely physical conception. Nor does it lead back to a pre-Marxist urban theory which abstracted from the social relations of production. I fully accept that activities relate to historically specific roles, social relations, institutions and mechanisms. The point, nevertheless, is firstly that given the physical characteristics of space (e.g. that two objects cannot occupy the same place at the same time) and the durability of the built environment, explanations are suspect which load all the responsibility for what happens on the particular social characteristics and coordinates of the activities and ignore the irreducible materiality of the activities and people themselves. Secondly, the very existence of the various activities depends on division of labour, though there is nothing asocial about that. Thus the extraordinary tangle of long journeys-to-work is a function of the social division of labour, regardless of the social relations of production. The vast array of specialized jobs and specialized workers cannot all be located in the same place; and since the vacancies and the people to fill them crop up at different times, it is common to find a worker with specialism x commuting miles to her job when there is an x-type job close to her home, for the mundane reason that the local job was already filled when she was seeking work. Similar such 'inefficiencies' occur in state-socialist cities and would occur were the worker seeking to join a cooperative in a market-socialist city. They are only removable in the fictional world of spaceless and timeless models of urban form where the built environment can be changed instantly and costlessly and every location and activity is continually up for auction (or, even more improbably, democratic vote).

Acknowledging processes such as these which are not unique to capitalist cities is compatible with recognizing those which are, such as their distinctive class relations. It is a matter of confronting the double nature of urbanization in advanced economies. However, urban political economy has generally failed to do this successfully.

One of the first attempts to use the distinction between the technical and social divisions of labour in explaining these matters was Lojkine's path-breaking essay on a Marxist theory of urbanization (Lojkine, 1976). He also confronted directly the issue of historical specificity. Previous, pre-Marxist work had taken urbanization to be an autonomous, transhistorical phenomenon. Though Lojkine criticizes this assumption, he does not merely reject it out of hand, but explains what gives rise to it:

> Urbanization has always been about the mobilization, production, appropriation and absorption of economic surpluses. To the degree that capitalism is but a special version of that, we can reasonably argue that the urban process has more universal meaning than the specific analysis of any particular mode of production.
>
> *(Lojkine, 1976, p. 124)*

and

> what does explain the apparent autonomy of urban phenomena is the fact that they are part of the *division of labour within society* and not the *division of labour within the productive unit*: now the 'social' division of labour – of which the division between town and country is the 'fundamental basis' – *is part of the economic formations of the most diverse societies and not, like the division within manufacture or that within the workshop, of the capitalist formation alone.*
>
> *(ibid., emphasis in the original)*

Leaving aside the questionable claim that the technical division of labour is unique to capitalism (it must feature in any advanced economy), Lojkine appears to do what we argued was required in claiming that a social process is historically specific, for he distinguishes those features of urbanization which are unique to capitalism from those which are not. He also discusses what is in effect the double nature of urbanization. By bringing different activities into close proximity, any city facilitates and speeds up interaction, thereby encouraging specialization or division of labour, but in capitalism the pressure for agglomeration occurs under its distinctive social relations, under the imperatives of capital accumulation and the drive to reduce circulation time, and in conditions in which land has been commodified.

Lojkine goes on to explain how these processes are at the same time obstructed by 'the laws of capitalist competition' and 'the parcelling out of urban space into independent fragments which are the private property of land-owners' (1976, p. 127). The latter 'contradiction' is especially important. Within a plot of land owned by a capitalist, activities can be rationally organized, but this becomes difficult where the activities are strung out across different sites, under different ownership, and where capitalists have to pay ground rent to land-owners to get access to land. From the standpoint of a mode of abstraction focusing on capital accumulation rather than the allocation of resources, rent is simply unproductive and a drain on surplus value. This contradiction or conflict between industrial capital and land ownership, together with property capital and its financial supports, is one of the most important features of capitalist urbanization (see especially Lamarche, 1976; Harvey, 1985). It is often internalized within individual capitals as they seek opportunistic advantages in both industrial production and property.

However, while Lojkine appears at points to recognize the double nature of the social division of labour and urbanization, he does not sustain the analysis. Instead of following through its implications and dealing with the extent to which 'contradictions' between activities derive from the social division of labour irrespective of the social relations of production – such that, for example, they could be expected under market socialism or state socialism – Lojkine reverts to an orthodox, one-sided accumulation perspective and to class reductionism. It is ultimately only capitalist relations of production which are held responsible for the tensions and contradictions of urbanization under capitalism.

Allen Scott's *The Urban Land Nexus and the State* makes much of capitalist urbanization's internal obstacles to rationalization (Scott, 1980). The analysis is primarily concerned with its anarchy, and more specifically with conflicts between private and collective action. Although Scott does not use the term 'double nature', in effect he provides a particularly full discussion of the issue in relation to urbanization, albeit one which I shall argue is flawed. The book is also notable for the extreme constructivism of its critical standpoint. For example:

> [T]he urban problem *par excellence* is the problem of how to control and rationalize in some reasonably definitive way, the overall pattern and dynamics of the urban land nexus. As it it, contemporary land-use patterns are ... the expression of an impersonal and (in human terms) irrational capitalist logic as mediated through urban space.
>
> *(1980, p. 65)*

further:

> although individual firms and households exert substantial private control over the development, exchange, and utilization of urban land, nevertheless ... none has control over the global outcomes that emerge from this circumstance.
>
> *(p. 137)*

Scott goes on to argue that the problem referred to in the second of the above quotations is

> so familiar a paradox of the contemporary urban scene that it is frequently taken in mainstream urban theory to be an immutable, universal feature of the urban process, and is reified into various versions of the decision-making under uncertainty predicament ... precisely because urban land development is privately controlled, the final outcomes of this process are necessarily and paradoxically out of control.
>
> *(p. 138)*

Here Scott is challenging the pre-Marxist treatment of urbanization which ignored the dominant mode of production under which it occurs. Although he is critical of accounts of urbanization which give it a single, transhistorical, technical nature, he is also aware of the converse fallacy when he criticizes 'orthodox Marxian theorists, with their insistence on the analytical preeminence of class struggle [who] generally overlook completely the technical hitches and encumbrances that precede all collective action in the urban land nexus' (pp. 142–3). While such comments indicate an awareness of the double nature of urbanization, its implications tend to be resisted rather than accepted. Many of the discussions fall into the category of technical explanations of inefficiencies and barriers to rationalization which are not inherently capitalist. Yet, like Lojkine, in insisting that the conclusion we should draw from such arguments is an indictment of the capitalist character of urbanization, he leans towards the second trap of orthodox Marxian theorists, where it has a single, class-based nature.

Throughout, there is an elision between 'private' in the sense of private individuals or organizations and 'private' in the sense of private capital. This is surprising, as Marxists are usually quick to point out that owning consumption goods like a house does not make one a capitalist. Externalities and conflicts between independent private owners are certainly a feature of urbanization but they don't have any unique connection with capitalist class relations.[13] Many urban conflicts arise from the attempts of different groups to exclude outsiders from their residential spaces (Davis, 1991). Even where it is property values that are at stake in such disputes, the links to capitalist social relations of production are tenuous. Scott argues that private, individualistic decision-making tends to be socially inefficient in both spatial and temporal terms, but he never demonstrates the link between the problems ensuing from the fragmentation of control and the capitalist control of some of the fragments.[14]

In a catallaxy, how *could* firms or households individually or collectively control the global outcomes? We may grant that some of the bad unintended consequences of individual actions that occur in our society could be foreseen and hence avoided. But Scott's contrast space, like that of Marx, is a society in which actions have not only no bad unintended consequences but no unintended consequences at all. The idea that conflicts between individual and social interests in urban development could and should be eliminated is not only constructivist but based on a

---

[13] As we saw in chapter 6, there is a certain irony in Marxist critiques of externalities, at least where they are made by those hostile to markets and private property. On the one hand externalities show the limits or failings of markets, while on the other, the removal of markets makes it difficult to assess benefits and costs at all. And at least markets allow the individual some control over what s/he consumes. Interestingly Hayek (1960) recognizes that externalities are unavoidable in urban areas and that consequently there is a role for state planning. In urban land and housing or urban means of transport, the removal of markets could intensify free-rider problems and result in overprovision of some goods, which always means underprovision of others.

[14] A discussion of the 'temporal myopia of private locational activity' (pp. 165–7), shows how successive individual locational decisions lead to inefficient social outcomes. While this is undoubtedly true, this is not uniquely connected to capitalist relations of production but is very much in the mould of the 'decision-making under uncertainty predicament'. It could be combined with an analysis of specifically capitalist processes, such as the speculative activity of property capital in locating offices or shopping malls, but it applies to any development process not under centralized control. At one point Scott effectively concedes this transhistorical dimension of urbanization; 'if it had been possible to accommodate urban activities within instantly convertible nondurable structures, there would in all probability have been many fewer problems in capitalist cities than is currently the case, and the collective undecidability of urban land uses would have posed quite negligible difficulties' (1980, p. 160).

view of the causes of those conflicts which never quite acknowledges the implications of the social division of labour and catallaxy (see also Harvey, 1985).[15] Conflicts between private and collective action are to be expected wherever there is a deep division of labour, even where the fragments of property are not owned by capitalists. The only way such conflicts could be avoided would be via centralized, authoritarian control which would (a) have to possess superhuman powers in being able to recognize, anticipate and rationally reconcile conflicts of interest, and (b) would allow little individual liberty and would have to be hostile to difference and value pluralism. Democratic negotiation and decision-making could of course be used for major developments, and to a greater extent than most capitalist societies already employ, but the information problems of extending it to the vast number of urban changes of social interest would be overwhelming. Moreover, unless there were limitations on voting, minorities would be at risk from tyranny-of-the-majority effects.

Harvey's critique is also highly constructivist, but there is a startlingly revealing moment in his essay on 'Money, time, space and the city' (in Harvey, 1985), where he briefly confronts his implied critical standpoint and considers alternatives to the anarchy of the social division of labour in the city. Referring to attempts of earlier generations of socialists to work out comprehensively planned 'allocations of space and time as opposed to those achieved by market processes in which money power played a dominant role', he comments:

> It took many years of bitter experience and reluctant self-criticism to recognize that the total rationalization of the use of space and time by some external authority was perhaps more repressive than chaotic market allocations ... Certainly to the degree that space and time are forms of social power, their control could all too easily degenerate into a replication of forms of class domination that the elimination of money power was supposed to abolish.
>
> *(1985, p. 32)*

In similar vein, Harvey notes that the

> nationalization of land and the abolition of property rights does not necessarily liberate space for popular appropriation. It can even lead to the erosion of those limited rights to appropriate space given by private

---

[15] In his later book, *Metropolis: From Division of Labour to Urban Form* (1988), Scott brings the division of labour to centre stage, but the book is primarily concerned with the spatial organization of industry, rather than the political economy of urbanization.

property and other mechanisms of securing social space. The prevention
of one mode of dominating space merely creates another.

(1985, p. 33)

This is of course the Weberian 'iron cage' scenario, though it is not
acknowledged as such. When confronted with irrationality, anarchy and
contingency, it is tempting to assume that rationalization will free people
from their effects and empower them. But while rationalization might
empower an individual or small group by bringing more of its environment under control, when generalized to a social level it has to enslave
people in order to eliminate cross-purposes and irrationality (Bauman,
1990). Fortunately, it is only partly feasible, but in any case, it is
certainly undesirable. Although Harvey does not pursue this line of
reasoning further, it suggests that in part he misrepresents capitalist
urbanization by attributing to capitalist social relations and economic
mechanisms effects which are co-determined by the intractibility of an
advanced social division of labour, and he effectively concedes our
argument in chapter 3 that the power of the working class is limited by
far more than its lack of ownership of the means of production.

   Although Harvey apparently recognizes the problems with this critical
standpoint, it has not led him to alter his explanations and critiques of
capitalism and money in subsequent writing (e.g. Harvey, 1989, 1993).
Marxists generally seem reluctant to take some rather obvious further
steps in the argument, perhaps because they conflict with orthodox
views, and raise the spectre of liberal accusations that Marxism has a
latent authoritarianism. For example, the bleak prospects of centralized
power, especially in the form of a major rationalization of the anarchy
of interaction in the city, invite us to judge that anarchy less negatively
than is customary in the Marxist tradition, where any kind of *ex post*
market regulation is resisted as involving alienation, economic forces
working behind our backs, and so on. It suggests that there is at least
something in the liberal argument that this anarchy also affords some
liberty, which – double-edged though it may be – helps to account for
the popularity of market allocation. (Ironically Marx also recognized
that the realm of exchange was or could be one of equality.) The fact
that liberals characteristically underestimate or choose to ignore the
inequalities, irrationalities, anarchy and uneven development of cities
does not detract from this point. Note, however, that such a concession
does not entail an acceptance of capitalism, for a market socialism in
our sense could avoid both capital and class and the iron-cage scenario.

   When the critical standpoints of radical urban political economy are
examined, the unrelenting negativism of its accounts (e.g. its condemnation of money, its evasion of the benefits of competition, rising

standards of living, life expectancy) looks all the more absurd. If not only state socialism but market socialism are unacceptable, what other kinds of economy are seen as desirable and feasible? On this we typically encounter silence or evasion. I assume this is because of an antipathy to markets and a fear that market socialism would retain too many unwanted features of capitalism, such as unemployment, inequality, macro-economic problems. This may be so, though the results would, as the literature on market socialism suggests, depend very much upon the actual mix of centralization and decentralization, market regulation, property relations, and so on (Bardhan and Roemer, 1992; Brus, 1972; Brus and Laski, 1989; Estrin, 1989; Nove, 1983; Horvat, 1982). A very different alternative is to reject advanced economies and return to small-scale local economies, so there is more scope for *ex ante* collective control, but as far as I know urban political economists did not have such an alternative in mind.

Within advanced economies, there could presumably be substantially more democratic control, more networking, more public provision and more planning of particular sectors than we currently have, but expectations that these could replace rather than supplement decentralized control and market coordination are vulnerable to liberal critiques such as Hayek's. As an example, let us return to the case of land nationalization and 'popular appropriation'. The socialist inclination is for land development to be controlled by the people, not the property developer. Here again the limitations of the policy suggest a problem with the diagnosis. Is the only obstacle to popular control of a city's land use the existence of capitalist interests in the shape of property capital? Or is it also that effective democratic control is infeasible due to uncertainty, information overload, division of knowledge, lack of common interests, free-rider problems and the complexity of what has to be controlled? (In a typical city there are likely to be thousands of decisions regarding location and land use each year, and each is likely to be highly specific to particular institutions, households or individuals.) It might very well be argued that individuals are often wrong in supposing that their decisions do not have any effect on others, and therefore that the decisions need to be 'socialized' more. But we cannot assume that many people care enough to learn about how their city works and become informed about particular planning decisions and to spend what would surely be a considerable amount of time participating in democratic decisions about them. Nor does it make sense to say that the achievement of popular control is just a problem of organization, whether political or 'merely technical', for in dismissing the intractability of what would have to be organized, this is a thoroughly voluntarist answer. I certainly don't wish to imply that we already have the best of possible worlds;

major property development decisions in capitalist cities are routinely made in outrageously undemocratic ways (CIS, 1973; Ambrose, 1987). I have no doubt that there is scope for increasing general participation in major planning and property development decisions, but beyond this the potential is unlikely to be dramatic. As the division of labour and the technological level of an economy develops, technocracy tends to displace democracy (Weber, 1947; Bobbio, 1987).

If effective popular control cannot be created simply by removing capitalist property development, then explanations of the problems wholly in terms of the power of property capital must be significantly incomplete. Marxist analyses consider property capital purely from a capital-accumulation and class perspective; that is, as a branch of capital; but it is also responsible for a specialized kind of work, i.e. it is part of the division of labour. Complex societies need to delegate much of the work to property specialists to plan and organize the construction of buildings and infrastructure, but they don't need to be capitalist organizations, able to appropriate development gain resulting from the actions of others, bloated with funds from financial capital, and with profit the only consideration.

Generalizing from these issues, we are again naturally led to wonder what 'popular appropriation' and 'socialization' could mean in the context of complex economies. Among small groups of self-selected people with similar interests, *ex ante* collective control can be highly liberating. Where there are vast numbers of diverse people with different interests, knowledge and material circumstances, the pursuit of collective *ex ante* control carries with it the threat of coercion and subordination to the plan. On the other hand, the anarchy of *ex post* control through markets has been associated with extraordinary degrees of uneven development, economic irrationality and poverty, with the freedom to lose as well as to gain. This is perhaps the central political-economic dilemma of an advanced economy, yet Marxism has had difficulty appreciating it because of its one-sided valuation of the relative merits of *ex ante* and *ex post* forms of economic coordination. Again, these arguments do not rule out the possibility of the abolition of class, or of the achievement of more democratic control over land use and the wider economy than we have in our existing society.

Thus radical urban political economy's critical analyses are made from standpoints that imply alternatives which are infeasible or undesirable, or both; they suggest a society which is thoroughly rationalized and yet somehow democratically controlled too.[16] Liberal concerns over how

---

[16] For Lojkine, the 'need for cooperation' is contradicted by capitalist competition and fragmented, private ownership of land (1976, p. 127). His critique is therefore implicitly made from the mythical standpoint of control of an advanced economy by the 'associated producers' (with the monopoly powers that implies).

much autonomy individuals and organizations should have are not even acknowledged. The oversocialized communist individual merges into the collective subject and difference and autonomy are seen as sources of irrationality. I strongly suspect that these illiberal implications are not intended by these authors, but they follow from their highly constructivist views.

Finally, I want to turn to a more recent and more openly normative work. From what might be termed an 'ultra-left' standpoint, Eisenschitz and Gough have developed a critique of local economic policy (1993) which has the merit of exposing, with great clarity, some of the dilemmas of the critical standpoints of radical urban theorists and policy-makers. Referring to radical, local economic policies in Britain in the eighties, they write:

> The Left ... was ambiguous about the basis of alternatives. On the one hand they were presented as anti-capitalist, meeting human needs irrespective of profitability; on the other hand they were said to increase productivity and profitability and therefore to be more rational paths for capital.
>
> *(p. 202)*

Eisenschitz and Gough identify a 'contradiction' between trying to defend and empower labour and improve working conditions – a goal evident in 'contract-compliance' policies of Left local economic strategies – and trying to modernize and increase the competitiveness of local industry. There *is* a dilemma here, or at least a tension between strengthening labour and making their firms more competitive, only it needs to be interpreted not only in terms of the 'vertical' relations of class and class struggle, but also in terms of the horizontal relations between different groups of workers. Other things being equal, in the short run and sometimes in the long run, raising one group of workers' incomes and lowering their work rates may disadvantage another group of workers or consumers by raising the prices of wage goods (or raising taxes to fund public-sector pay increases), and this is possible in any advanced economy, be it capitalist, market socialist or state socialist. The trade-offs therefore run not only between capital and labour or profits and wages, but between different groups of workers or consumers. Of course, in practice, other things are rarely equal, but that does not mean this conflict can always be resolved. For example, it is not impossible under capitalism for firms simultaneously to produce higher-quality goods at lower price, pay higher wages and offset the cost through increased productivity and output, hence avoiding redundancies, while introducing new technology that raises output per worker *and* reduces stress on workers. But while this is possible it is likely to be rare, given the difficulty of producing this happy conjunction of events and the fact that production is controlled in the interests of capitalists. However, these dilemmas are not restricted to capitalism; even in a market-socialist economy, there are likely to be cases where wages can only be raised or work conditions improved at a higher cost to consumers or taxpayers, given the difficulty of raising productivity in some kinds of work. (This may of course be a price worth paying.) Even if it were possible to have a moneyless advanced economy, giving more to one group would generally mean giving less to others. Political economy has to consider not only vertical relations of exploitation but horizontal relations of distribution and allocational efficiency.

For Eisenschitz and Gough, 'the over-riding aims of local socialist economic policy should be political: to strengthen collective organisation and solidarity and combat divisions within labour' (p. 208). They appear to have no economic alternative save that of international democratic control! They ignore the inherent limitations of democratic control as a means of mirco-economic coordination, and say nothing about *how* economies are to be regulated and scarce resources allocated among competing ends. Aside from the question of whether 'combatting' the divison of labour means reducing it – and hence almost certainly lowering levels of development – if the Left were successful in strengthening collective organization, it would then have to face the dilemmas summarized in figure 3.3. Political economy – inside and outside urban and regional studies – has to address issues of division of knowledge and catallaxy, horizontal control, coordination, allocation and economizing, which generally *elude* democratic control (Nove, 1983; Sayer and Walker, 1992). To ignore these issues is simply to duck the most serious problems of political economy and opt for infeasible constructivism. In short, while Eisenschitz and Gough confront the dilemmas of radical urban analyses and policies, they exaggerate their peculiarity to capitalism and evade the incoherence of their own alternatives.

Conclusions

I have argued that we need to re-examine political-economic theory and its application to urban and regional development. The major strides in radical theory made in this field in the 1970s and early 1980s can neither simply be accepted as adequate or forgotten. Rather, the theory needs to be re-evaluated systematically so that we can decide what to retain, what to change and what to drop.

Radical urban and regional political economy aspires to be a critical social science, but its critical standpoints are curiously under-examined. Its constructivist character, coupled with its reluctance to ask counterfactual questions, reinforces the tendency to reduce social phenomena to a single capitalist nature, by assuming that whatever occurs within capitalism must be a unique product of the diagnostic features of capitalism. Often there are specifically capitalist *variants* of dilemmas and problems that would occur in other advanced economies. As we have seen, some theorists of urban and regional political economy attempted to grasp the double nature of the state and of capitalist urban development, but without success. Both to explain urban and regional change and to criticize and evaluate it, we need to examine not only what is uniquely capitalist about it but what is *not* uniquely capitalist about it. Ditto the state and industry. Once this double nature is recognized, the dilemmas or problems of urban and regional development are cast in a new light – sometimes as genuinely unique to capitalism, sometimes as shared by all market economies, and sometimes – as in the case of the division of knowledge and its effects – shared (albeit in somewhat different forms) by both market and centrally planned advanced economies.

The dilemmas arising from the tensions between anarchy, rationalization and liberty can't be eliminated by changing the social relations of production. This suggests that a moderate form of the industrial-society thesis could profitably be reconsidered in urban studies: while the particular social organization of production makes a significant difference to industrialization and urbanization, it does not make all the difference. Theorists therefore need to reconsider not only how far their critiques of urbanism are critiques of capitalism and how far they are critiques of modernity (Savage and Warde, 1993), but how far they imply a critique of industrialism. Early sociologists who wrote on the city, such as Simmel and Tönnies, attached great significance to division of labour and its implications, though without analysing this in terms of political economy. Their profound ambivalence about modernity and urbanism can be taken as responses to their dilemmatic qualities. Such theorists were brusquely rejected by the new Marxist urban theorists such as Castells for allegedly having mystified the capitalist nature of the city (Castells, 1976). Although, as Savage and Warde argue, this was not entirely fair, the early urban sociologists were also legitimately addressing developments which were not entirely specific to capitalism. Consequently, while we need to problematize the relations between capitalism, urbanism and modernity, as they suggest, we also need to consider the relations between capitalism and industrialism (see also Giddens, 1985). The new radical urban studies literature of the 1970s and early 1980s corrected the neglect of political economy but overlooked industrialism in its determination to make class the pre-eminent form of social division.

Since that time, the class reductionism of this literature has been corrected in so far as gender and race are concerned, but the significance of the social division of labour or catallaxy as a source of division and a key feature of urban and regional development is still not adequately appreciated. We have still to get the balance right.

9

Implications

Rather than summarize the conclusions of the foregoing chapters I want to follow up some implications of the main themes of the book.

Our evaluation of radical political economy has in part been jointly conducted with a critique of liberal or mainstream economics. These adopt radically different 'modes of abstraction', each with its own systematic biases and absences. Yet in many cases there are complementary strengths and weaknesses. I now want to draw out these tensions and relationships, so as to suggest how we might move beyond them. I shall then assess their strikingly different conceptions of the nature and purpose of political economy or economics. This is often discussed in terms of Polanyi's distinction between substantive and formal approaches, the former focusing on the shape and functioning of economies, the latter on economizing or choice. Radical political economy has generally allied itself exclusively to the substantive definition. I shall argue that even if one's interests are substantive, economizing cannot be neglected.

'Double-nature' arguments have figured recurrently in our critique, and at several points I have suggested that radical political economy needs a moderate version of the industrial-society thesis, so as to acknowledge the extent to which advanced economies are shaped by their industrial, as well as their capitalist or socialist character. Concepts of industrialism have a long history in social theory and are bound to be contentious in the current context, since the revival of radical political economy in the 1960s was partly a reaction to them. In view of this I want to distinguish my own version of the industrial-society thesis from earlier ones and position it in relation to the the work of theorists such as Weber and Foucault which includes some convergent arguments.

We noted at the outset the peculiarities of the intellectual history of the period since the late seventies, with postmodernism rising in academe as Marxism faded, while liberalism – often hardly acknowledged in radical academe – was dominant in politics outside. Here I want to connect postmodernism to the battles between radical political economy and its old enemy, liberalism, by drawing attention to some striking similarities between postmodernism and liberalism.

Our critique has not only been addressed to the substantive claims of radical political economy but to its implicit view of the nature of critical social science. In particular we have argued that critical standpoints need to be examined. Here we take this further and argue that if radical political economy is to have much emancipatory potential it must develop links with normative political theory. The need for this has become all too clear with the loss of confidence on the Left regarding alternatives. I then provide a sketch of the main contending political philosophies and theories that might be drawn upon in rethinking the critical standpoints of radical political economy. Despite the many critiques of liberalism from various quarters, it has some advantages in the context of the rise of multi-ethnic societies within a global division of labour, which the Left cannot afford to continue to ignore.

On the Relationship between Radical Political Economy and Mainstream Economics

Cut across a log, and you reveal a pattern of concentric rings: cut along the length of the same log, and the grains forms a pattern of parallel lines. The two patterns could hardly be more different and yet they are part of the same structure. If you look at capitalism from a Marxist standpoint you find capital accumulation and class: if you look at it from the standpoint of liberal economics you see exchange and economic calculation regarding the allocation of resources among competing ends. It is of course customary to think of these two approaches or modes of abstraction as mutually exclusive and contradictory, and there are indeed issues on which they flatly contradict one another, such as the nature of wages and the origin of profit, and for which the log analogy does not apply. However, the arguments of the foregoing chapters imply that in some respects the concerns of the two modes of abstraction are no more contradictory than the concentric rings and the parallel lines of the wood grain; they are radically different but complementary representations of different aspects of the structure of market economies. Just as one would understand little about the structure of wood from making cuts in just one direction, so one misses a great deal by ignoring either of these two kinds of abstraction. Yet the two approaches to the study of economies are largely blind to each other's concerns.

So in radical political economy – both in its accounts and its critical standpoints – we find a consistent lack of interest in the allocational properties of economies and a tendency to transpose the effects of allocational mechanisms into consequences of class. Thus in relation to capitalism, the effects of markets are transposed – with more faith than justification – into outcomes of capitalist social relations of production (e.g. Clarke, 1982; McNally, 1993); indeed, making such a transposition was a way of establishing one's Marxist credentials. Similarly, the negative effects of central planning in state socialism are blamed on its (quasi-)class relations.

Marxist and liberal modes of abstraction in economics are therefore orthogonal to one another. Where liberal economists treat production (supply), distribution, exchange and consumption (demand) as separate elements, Marxists, following Marx in the *Grundrisse*, tend to integrate these as 'moments' in a dialectical movement dominated by production. Thus distribution is first of all distribution of the means of production and hence not something subsequent to production, consumption is not autonomous because consumers are also workers and consume to reproduce their labour power for production, and production involves productive consumption, etc. (Marx, 1973, pp. 88–100. This form of abstraction does have the virtue of showing the interconnectedness of these moments, relating them to class and confronting the way in which economic systems are reproduced, which liberal economics largely misses. But it achieves these insights at the cost of burying the problem of micro-economic coordination of the division of labour and allocation of resources. This is not to say that Marxism takes the reproduction of the system to be automatic; indeed the risks of non-completion of circuits and hence failure to realize value are often noted, but this still skirts the problem of allocation and coordination. Nor does Marxism wholly ignore issues of efficiency, but it is only concerned with it in terms of reduction of socially necessary labour time for the production of particular commodities, not with allocative efficiency across substitutable commodities when choices must be made between alternatives.

Liberals are classically concerned primarily with consumers and take little interest in production and its social organization. Marxists prioritize producers, marginalizing consumption or assimilating it into production, as in 'the reproduction of labour power'. I see no contradiction in acknowledging the liberal point, deriving from Smith, that the whole point of economic activity is to satisfy consumption; and in insisting, with Marxism, that production is a condition of the existence of social life, and that its particular form of social organization affects the rest of society. The liberal view misses the materialist insight, but Marxism's production optic can also lead to absurdities; for instance in the many studies of the labour process that followed the publication of Braverman's *Labor and Monopoly Capital* (1974), one could be forgiven for imagining that the point of producing cars or whatever (usually cars!) was as a means towards the goal of dominating labour rather than as a way of meeting the demand for cars.

The complementary insights and blindspots are strikingly evident in the way in which Marxists see the mobility of capital in terms of the power of capital over labour, while liberals treat it in terms of efficient and neutral responses to ever-shifting patterns of costs and demand. Marxism's general dismissal of allocational issues is complemented by a hostility to the specific form in which they are determined in capitalism – that is, on the basis of profitability in market operations. The possibility of exit – of consumers changing to different suppliers – is seen only in terms of how it impacts on workers, and its consequences interpreted in terms of the actions of capital dominating labour. Marxism therefore ignores the benefits of choice as exit – both because it has no interest in what benefits people as consumers, and because it refuses to see that the losses for certain workers would be complemented by gains for other producers and consumers, though of course there is no reason why any sort of balance should result from the restructuring. Equally, liberalism is extraordinarily negligent in addressing the inequalities and inequities that derive from the mobility of capital.

Similarly, I see no contradiction in acknowledging that catallaxy and class are both important; indeed the former is vital for understanding the fragmented nature of all classes in capitalist society and why the fragmentation is unlikely to disappear in any advanced post-capitalist society.[1] Private ownership – particularly of the means of production – gives power over those lacking it (a power which I find illegitimate), but with an advanced division of labour it also brings power and responsibility into more correspondence than is possible in most cases of 'social ownership', though *less* than in the case of what we have termed 'restricted collective ownership' (chapter 6). Understanding ownership and control requires a simultaneous analysis of the vertical relations emphasized by Marxism and the horizontal relations emphasized by liberalism.

Radicals tend to see inequalities as deriving primarily from relations of domination, such as those of patriarchy or capital and labour, while liberals tend to see inequalities as resulting from differences in merit, misfortune, or from the unintended consequences of actions. In the latter case, markets are the paradigm case; winning or losing in markets is not

---

[1] As Meszaros has noted, the explanation which Marx offered for the lack of revolutionary potential of peasants – by virtue of their fragmentation – applies equally to the proletariat (Meszaros, 1987).

dependent on any intrinsic merit but on how far one satisfies the demands of others, and inequalities can open up as indirect consequences of the distant actions of unknown others. Liberals are often blind to relations of domination because of their individualistic representation of action, which treats all social relations as external and contingent, and which has difficulty recognizing anything between free choice and coercion (Holton, 1992, p. 215). But relations of domination are not the only source of inequality, and assuming they are is particularly problematic in dealing with a market economy and its myriad indirect connections. Moreover the two sources of inequality can feed into one another; as we saw in chapter 7, initially equal petty producers who differ in their success in finding buyers for what they sell may end up as capitalists and wage labourers, thereby entering into a relation of domination, while those already in such a position are likely to find their inequalities reinforced by the unintended consequences of actions in markets. We can also recognize that many 'choices' or actions are pressured and undertaken under duress even if they are not directly coerced. Once again, the liberal and radical views have complementary blindspots.

I realize my rustic analogy of the log and the wood grain will jar with the dominant view according to which radical political economy and liberal/mainstream economics are simply different, incommensurable, discourses/problematics/paradigms. However, as a general point about methodology, I would question the enthusiasm for such judgements. Adherents of these different discourses share the same life-world, their discourses overlap. Where they contradict one another they must have something in common, over which they contradict – otherwise they are merely at cross purposes (Sayer, 1992a, pp. 72–9). Where they are incommensurable, this can only matter if the relevant concepts concern the same subjects; no one worries that football and tennis are incommensurable (Bhaskar, 1989). Marxism and subsequent radical political economy have been forged through historical engagements with rival theories – with Scottish political economy, Hegelianism, and later marginalism. Marxism is not at odds with or different from these in every aspect, and where it is, they may have something in common over which they are at odds. Exaggerating the incommensurability of rival theories provides a convenient excuse for ignoring external criticism. The concepts of political economy and mainstream economics may differ, even where they share the same terms, but it is not impossible to understand both simultaneously, even if it is not always possible to accept both simultaneously without contradiction. Some of their differences derive from differences in their objects or the tasks they set themselves, and here, there may be scope for combining their insights. Here the social division of labour functions as a 'hinge concept' – the more one contemplates its materiality and associated division of knowledge, the more one is forced to move beyond Marxism and to consider liberal views concerning the 'horizontal' relations among producers and consumers.

I recognize that in some quarters such arguments are likely to provoke charges of eclecticism. The idea of combining the good parts of each theory and leaving out the bad sounds naïve and is liable to raise sneers. Usually modifications are required before elements of different theories can be combined, and we have noted the need for these repeatedly. Moreover, we have not attempted to deny that there are many issues on which they cannot be reconciled. However, one thing is clear – the question of the compatibility or otherwise of different theories cannot be an *a priori* matter, or one that is answered on the basis of extraordinary views of different theories as constituting wholly separate discourses, coupled with the prejudice that what is different must be incompatible. To refuse even to consider the possibility of combining elements (suitably modified) of different theories is just dogmatism.

Defining the Economic Problem: 'Substantive' and 'Formal' Conceptions

> [In Marx] . . . the question of economic calculation (in its most general sense as the criteria and methods for choosing between different uses of the available means (resources) for the achievement of optimum economic results) was never, or hardly ever, the subject of consideration.
>
> *(Brus, 1972, p. 28)*

Political economy and liberal or mainstream economics differ in how they define their object of study. Following Polanyi (1957) it is common to distinguish political economy from mainstream economics on the grounds that the former is concerned with 'substantive' economies and how they work, whereas the latter is concerned with the *formal* meaning of 'economic', involving choice or economizing among means to ends under conditions of scarcity. Most radical political economists, with the unsurprising exception of 'analytical' or 'neoclassical Marxists', would appear to accept Polanyi's view that the substantive conception should be the dominant one, for they generally ignore issues of economizing.

Polanyi argues that while the formal conception assumes that there is a transhistorical problem it is actually specific to capitalist societies. In capitalist society it is necessary to compare costs, and economic practices involve, as a condition of economic survival, choice among competing means to ends. But in other societies, particularly pre-modern ones, economic practices are matters of convention; where trade occurs it is highly regulated so that prices and other aspects of exchange are fixed. Scarcity is said to be relative, and in situations of natural abundance and limited, socially defined needs, not a problem at all, with the result that people in pre-capitalist societies may need to work less than workers in modern economies. Pre-capitalist economic behaviour is strongly embedded in other practices and actors follow conventions, rather than make rational choices, though conventions may provide quite rational solutions to problems. By contrast, markets have the effect of disembedding economic relations to some degree.[2]

The danger of a transhistorical definition of the economic problem is that it can lead one to overlook the historically specific ways in which different societies define their economic problems or tasks and to project capitalist priorities onto non-capitalist societies. While Polanyi's criticisms of ahistorical applications of formal conceptions of the economic are important his argument about the proper object of study is less coherent. Although he concedes that the formal definition is appropriate for capitalism he argues that 'only the substantive meaning of "economic" is capable of yielding the concepts that are required by the social sciences for an investigation of all the empirical economies of the past *and present*' (Polanyi, 1957, p. 30; emphasis added). While there are some grounds for resisting the formal conception of economics in the study of pre-capitalist economies, this does not justify ignoring it in capitalism where it has most credibility.

In practice, the distinction between substantive and formal corresponds to that between empirical studies of economies and normative studies of how self-interested rational actors should respond to alternative possibilities under often highly implausible and idealized conditions. It is the latter which (notoriously) carries most prestige in mainstream economics, though economists like to pretend that the latter is actually a positive kind of inquiry capable of illuminating what happens in practice. While I have no wish to endorse the mainstream version of the formal conception, I shall argue that there is a limited version of the transhistorical definition that is acceptable, although mainstream economics characteristically misuses it; and secondly that radical political economy cannot afford to exclude the formal considerations if it is to understand substantive economies.

Our preferred version of the 'formal' conception of economics is based on the fact that resources, including labour power, always have alternative uses, and that choosing one use precludes others, if only because we cannot do many things simultaneously and time is always limited. The

---

[2] Actually, the extent to which this contrast between embedded pre-capitalist practices and disembedded market processes holds is contentious. See Granovetter, 1990.

solution to this problem is usually considered in terms of maximizing utility/minimizing costs (where costs do not necessarily have to be assessed in money terms). Each kind of society has its own way of meeting its needs and, whether their members realize it or not, a way of dealing with this transhistorical problem of economizing. In self-sufficient pre-capitalist agrarian societies, people know very well that if they do not bother to sow crops, weed or irrigate their fields they will go hungry. Their economic practices may be deeply socially embedded, convention-alized and stable, but this does not mean they are totally indifferent to opportunity costs and that they can escape the materialist preconditions of social life, indeed these are far more visible than in more 'advanced' economies. The fact that such practices may be highly stable compared to those under capitalism does not mean that there are no alternatives to them, merely that they are not pursued. Such behaviour should not be seen as necessarily irrational: on the contrary, as development economists and agronomists have slowly learned to their cost and at the expense of those they were supposed to help, folk practices and conventions – for example concerning crop rotations and combinations – are often a good deal more successful than imported ones based on supposedly rational choices among alternatives (Chambers, 1983).

The resolution of the transhistorical economic problem is therefore not necessarily a system goal or a goal of individuals; it is usually no more than an unacknowledged or partly acknowledged condition, and also a largely unintended consequence, of actions motivated by more limited considerations. That economic actors may be oblivious to economizing options does not mean that alternatives count for nothing, for failure to take advantage of them still produces allocational effects which react back upon their subsequent actions, enabling and constraining them in various ways. Issues of economizing therefore always involve alternatives or potentials which are currently unrealized. The formal concept of economics should be treated as no more than a description of a transhis-torical *predicament* facing all societies: only in some societies are actions continually reassessed and changed in response to that predicament, but it is still has implications for those which don't recognize it.[3]

---

[3] This concern with what could be might initially seem a strange preoccupation for any social science, but all sciences have to understand not only what their objects actually happen to be doing at any moment in time, but what they are capable of doing (Bhaskar, 1975, 1989; Sayer, 1992a). This is so not only in the obvious case of the concept of market choices. The concept of unemployment has no meaning unless we can distinguish what individuals are potentially capable of doing from what they happen to be doing now. The crucial Marxist distinction between labour power and labour depends on the identification of potentials even where they are not exercised. Theories of gender as something which is socially constructed depend on acknowledging that men and women are capable of doing other things from what they happen to be doing now.

In pre-capitalist societies where there are mostly self-sufficient economic units, the producers benefit or suffer according to how they allocate their resources among alternative (competing) ends. Here division of labour is limited and producers control their own division of tasks *ex ante*. This is clearly different from economies with a deep division of labour and competitive markets, for instead of controlling the division of tasks themselves, individuals, enterprises and sometimes whole regions or countries occupy niches within a wider division of labour which they do *not* control. The hitherto separate problems of economizing in particular places coalesce and escape the control of any particular group, becoming a process 'working behind their backs', *ex post*. Economizing now takes the form of competing for customers and resources, and the fate of individuals, enterprises or regions is bound up with particular kinds of work from which it is usually difficult to change. The effects of markets in producing allocational efficiency on a social scale may seem impressive in the abstract, and so too might the effects of competition on productive efficiency, but they also mean that particular producers are no longer in control of their own destinies. Marketization strategies are therefore highly risky. This horizontal displacement of control, along with the horizontal displacement of people's relationship with nature, make the development of divisions of labour an extraordinarily fateful process in human history. Our interests as producers and as consumers are divorced from one another and ecological constraints are no longer local and pressing.

In state-socialist societies, the problem of economizing takes the form of a continual, conscious assessment of alternative uses of resources; competition between different projects and different ways of using resources is always present and is mediated and resolved politically.[4] Again, in so far as people occupy particular locations within divisions of labour, they have a stake in the outcome of this politicized competition though they cannot take unilateral action to defend their situation by competing, as their equivalents in market economies can, but then nor do they risk unemployment or hunger if their work is uncompetitive. Where there are many enterprises producing the same product, as in the

---

[4] Curiously, while Polanyi discusses the Soviet system as involving redistribution, he fails to note that it also involved choices among alternatives. There are many further questionable points in Polanyi's essay, such as his assertion that choice is induced by insufficiency of means rather than also by the existence of alternatives. He also wrongly limits economizing to the choice of means to ends, ignoring its role in the selection of ends, for there are usually trade-offs between these too. Moreover the limitation of rationality to the formal kind, as if substantive ends could not be the subject of any kind of rational consideration, reproduces rather than challenges liberalism's agnosticism about ends.

case of coal mines, differences in productivity and quality of output invite economizing, favouring those producers with the highest and best yield per unit of input, as we saw in the Vorkuta case in chapter 8. Compared to capitalist societies, however, shifts in opportunity costs are harder to detect and there is less pressure to act upon them. However, again, from a critical standpoint, opportunities that are not acted upon can't be ignored, for they represent forgone development opportunities and hence help explain lack of development. While it is normal in radical political economy to criticize the creative destruction of capitalist uneven development, it is less common to criticize allocational inefficiency, and the opportunities forgone as a result. This is not to idealize the unfettered pursuit of the most 'competitive' possibilities, for this usually involves consequential losses, which are not borne by those taking the decisions. Active competition is restricted to particular kinds of economy, and it has destructive as well as beneficial effects, but it is in the nature of the transhistorical economic predicament itself that there are always trade-offs, whether they are acknowledged and weighed or not. To note that societies differ in the extent to which they assess these is not to endorse the form any assessments may take and the way decisions are made regarding them in particular societies. The economic concerns, activities and mechanisms of different societies may be strikingly different, but they all economize somehow. We can recognize the latter without denying the former.

So in accepting a qualified role for a conception of the economic problem as involving economizing in this way, I am certainly not condoning the common practice of mainstream economists of projecting capitalist rationality back onto pre-capitalist societies. As we saw in chapter 4, this usually involves a slide from acknowledgements of opportunity costs to the notion of latent markets and thence to imagining that markets actually exist; from 'in the beginning were markets' (Arrow, 1974) to 'markets are everywhere'. If, thanks to the exchange optic, one imagines (wrongly) that economic choices among alternatives must involve markets, then it can appear that better kinds of economizing can only come by developing markets. This is the egregious error of the World Bank prescriptions for marketization identified by Mackintosh that we noted earlier (Mackintosh, 1990).

Radical political economy's lack of concern in economizing is some-times defended on the grounds that political economy just isn't interested in questions of demand and supply, and the allocation of resources; rather it is interested in the broad structures or forms of social organization of economies, particularly the way in which production is organized. However, not every issue can be avoided by saying one simply isn't interested in it, especially if it has significant implications for what one *is* concerned with. This is the case with economizing. The shape and development of economies depends on how actors and institutions economize, not only within particular organizations, but across different organizations. I don't wish to abandon the idea that political economy should be primarily about economies (the 'substantive' definition), but it has to take account of economizing if it is to achieve its primary aim. It is particularly ironic that those who believe that a socialist economy is or should be one in which resources are allocated rationally on the basis of need, should refuse to take any interest in how resources are allocated in actual and potential economies.

The disregard of economizing relates to the neglect of formal rationality in normative radical thought. It is one thing to criticize markets and their power, but another to complain about some activity being subject to cost constraints. As Tomlinson (1986, p. 140) notes, there is a general reluctance on the Left to discuss allocative efficiency, intra-class priorities, divergent objectives and trade-offs, as if any talk of such issues were simply a cover for cuts, and as if political will were the answer to everything. Every economically useful thing has a cost, in the sense of possible alternatives forgone. The slogan 'each according to their need' is inadequate because it ignores scarcity and merit. Moreover it is ecologically unsound to ignore economizing. In this regard it is of course richly ironic to associate capitalism with economizing, given its voracious appetite for growth and unprecedented exploitation and destruction of nature – though it has economized especially on labour time per unit of output. The appropriate response to this irony is not to throw out concerns with economizing, but to seek better concepts of economizing, ones which conserve ecosystems rather than encouraging their destruction (Altvater, 1993).

Further objections to allowing a place for an interest in economizing might be directed at the usual baggage that we associate with such concerns in mainstream economics – the exchange optic, the neglect of production and innovation and an undersocialized view of economic action. But we can admit the formal problem without taking these on board since economizing and choice need not presuppose exchange; we are therefore not legitimizing the notion of economics as the study of exchange (Ferber and Nelson, 1993). Nor do we have to follow mainstream economics in marginalizing production or making it appear to involve no more than a choice between different combinations of inputs and outputs. Nor need a conception of economizing presuppose an undersocialized conception of economic action, as if it were a matter of isolated individuals making choices in a vacuum.[5]

---

[5] Equally, as we saw in chapter 5, socialization does not rule out choice, as if people were

I accept that political economy is properly primarily interested in substantive economies, but economizing cannot be ignored if we are to understand them: to use those wonderful old 'p-words' of Adam Smith, 'parsimony', 'prudence' and avoidance of 'profligacy' are still important.

Double-Nature Arguments and the Industrial-Society Thesis

We have already recommended the idea of a weak version of the industrial-society thesis. Lest this is misunderstood, I want to distinguish this from stronger and more familiar versions of the thesis. These would claim that the main features of highly developed economies derive not from the social relations or mode of production but from the nature of industrialization itself, such that it makes little difference whether the economies are capitalist or socialist: as long as they are industrial societies, they will have certain characteristics – widespread use of money, a high level of formal rationality, hierarchy, bureaucracy, markets, etc. (Badham, 1986). Corresponding to this are accounts or theories of urbanization which stress features such as the density of interaction, the functional nature of relationships, and so on, paying little or no attention to the social relations of production or treating them as natural accompaniments of industrialization and urbanization (e.g. Wirth, 1938). Both attach importance to division of labour, often more than to class. Neither of these presented a coherent *political-economic* theory, as Marxism does. They involved often vivid narrative accounts and ideal types, enumerations of processes and conditions, but with little in the way of analysis of structures and tendencies or development processes to compare with Marxism.

The danger of these theoretical tendencies is of course that they can be allowed to naturalize and conceal the specifically capitalist forms of society behind the bland exteriors of industrialism and urbanism. This reductionism did not go unchallenged, however, and some theorists attempted to negotiate between concepts of industrialism and capitalism. Dahrendorf sought to identify 'those factors which can be shown to be generated by the industrial production, and which cannot disappear, therefore, unless industry itself disappears' (1967, p. 40). Likewise Bell attempted another analytic separation: 'the terms feudalism, capitalism, socialism are a sequence of conceptual schemes, in the Marxist frame-

---

mere dupes; indeed it is hard to imagine how individuals and organizations could make choices without being socialized. For example, in order to respond to peer-group pressure we often have to choose quite carefully what we do, how we work and consume, unlearning previous behaviour patterns.

work along the axis of property relations. The terms pre-industrial, industrial and post-industrial are conceptual sequences along the axis of production and the kinds of knowledge that are used' (1974, p. 11). One of the main questions of the old industrialism/capitalism debate was therefore 'how far the modern world is a result of the expansion of capitalism or how far alternatively it is the outcome of the spread of industrialism?' (Giddens, 1987). In my view this remains of critical importance.

The dominant position in the debate has shifted several times in the last forty years. Having once been an orthodoxy, the industrial-society thesis came under heavy fire in the 1970s and 1980s with the resurgence of Marxist political economy; industrialism as a category was rejected as both redundant and ideological (Bottomore, 1985; Scase 1989). Moreover, where the largely sociological approach of the industrial-society thesis glossed over the analysis of economic mechanisms, the Marxist revival made these central, thereby greatly advancing understanding of industrial and urban development. At the same time, Marxism was influential in the development of feminism and cultural studies, which took radical social science beyond its narrow economism. By the end of the mid-1980s the tide began to turn and certain post-Marxist theorists began to argue that a concept of industrialism was needed *in addition to* capitalism and socialism, though equivalent arguments have still to be made regarding urbanization (e.g. Giddens, 1981, 1985).

These shifts can be related to our 'double-nature' theme. The strong version of the industrial-society thesis treats its object as having a single nature deriving from its technical or material character. The Marxist critique often tended towards a converse emphasis, with authors appearing to feel compelled to demonstrate that features such as inequality, conflicts between formal and substantive rationality, hierarchy and discipline were overwhelmingly consequences of the capitalist character of society, thereby implying that a socialist industrial society could be free of such features (e.g. Marglin, 1974; Braverman, 1974). Not only large-scale industry, but the social division of labour, urbanization and the state have a double nature. While Marx himself invoked the double-nature thesis occasionally, Marxists tend to have paid it less attention. Marxism's emphasis on internal relations (Ollman, 1971) has the effect of exaggerating the uniqueness of the interconnections found in any society. This is compounded by a tendency to neglect comparative and counterfactual inquiries, which could reveal the extent to which certain features of industrial society could be independent of capitalist forms.

Our position lies between the strong-industrialism thesis and the Marxist reaction. It gives more emphasis to an advanced division of labour than to the trappings of industrialization in terms of machinery, use of fossil fuels, etc., and it does not underplay the the role of class and other social relations. While I have argued that *some* of the features often seen as unique to capitalism are common to other market societies or to all kinds of industrial society, I obviously accept that the differences between capitalism, state socialism and market socialism are too great to ignore. There is therefore no question of choosing between industrialism and capitalism (state socialism, etc.): both concepts are needed.

Certain other strands within recent and classical social theory are convergent with this way of thinking. By the late 1980s, the work of Foucault had begun to make an impact on radical social science. His analysis of the development of disciplinary power and surveillance, of the many ways of regulating, integrating and disciplining activities in space–time, does not invoke any specifically capitalist causes. Technical rationality itself is a form of domination which capitalism did not originate – in fact it seems that early capitalism took advantage of the technologies of power already developed in non-capitalist institutions such as workhouses and prisons and modelled its technical divisions of labour on them (Foucault, 1977; Bauman, 1982). These ideas therefore have the effect of weakening the privileged link between capital and the forms of discipline and power within the capitalist firm.[6] In *Discipline and Punish* (1977), Foucault switches without remark from examples of factories to Jesuit colleges and to elementary education, noting the similarities in their micro-geographies, their partitioning of space and regulation of bodies. Discipline is analysed as a technique for constituting individuals – 'discipline fixes; it arrests or regulates movements; it clears up confusion; it dissipates compact groupings of individuals wandering about the country in unpredictable ways; it establishes calculated distributions' (1977, p. 219). He also challenges one-sidedly negative views of power, noting its role in constituting social objects, and exclusively 'top-down' views, emphasizing by contrast its 'capillary' nature, operating in the minutiae of everyday behaviour (1980).

As Foucault himself hints (1980), his views of power are not necessarily incompatible with more conventional associations with relations of domination. His work can therefore be interpreted as convergent with the double-nature theme. Thus, Foucault's work helps to support a more critical view of power within the state, one that countered the tendency to think of it as a powerless medium (or 'propertyless medium' as Habermas puts it), merely taking on and transmitting the priorities of

[6] Compare this with Marglin's much-cited essay in which workshops and factories are said to have been developed by nascent capital, not to use large-scale machinery but to enable the disciplining of workers (Marglin, 1974). The two accounts might not be incompatible; Marglin's evidence could be interpreted as identifying capitalist appropriation of existing technologies of power not previously used for economic production.

capital. As we argued in chapter 8, the state is not only a *capitalist* state: it is also a *state* whose properties are not reducible to effects of capitalist society. None of this would justify abandoning political economy for a Foucauldian approach, for the latter lacks specifically economic concepts, but it provides a useful supplement.

An older source of concepts of industrialism, but one which has experienced a revival of interest, is the work of Weber (1968 [1922]). His account of economy and society in effect substitutes 'industrial society' for 'capitalism' and 'socialism', emphasizing not capital and labour and appropriation but a society characterized by large-scale industrial production, technical rationality, the inexorable power of material goods, bureaucratic administration, and a pervasive 'calculating attitude' (Bottomore, 1985, p. 26). Given the vast literature evaluating Weber, I shall restrict myself to brief comments on the relationship of his thought to radical political economy and liberal thought. In Marxist circles Weber is often known primarily in relation to his concepts of class and social stratification, which are seen as rivals to Marxist class theory. It is here, for example, that Weberian influences have had most impact in radical urban and regional political economy (e.g. Saunders, 1983). However, in my view, Weber's theory of rationalization has far more profound critical implications for radical political economy.

As Derek Sayer and others have argued, while Weber posed powerful challenges to radical views of capitalism and socialism, it is important to note the similarities between Marx and Weber. Both saw the outcome of the process of industrialization in terms of the subordination of human beings to things, and of the producer to the product (D. Sayer, 1991; Bottomore, 1985), only in Marx, it is capitalist society rather than industrialisation *per se*, or rationalization. Equally the similarities and tensions between Weber and liberalism are richly complex (Bellamy, 1992). Thus while he shared the liberal fear of large-scale rationalization of society, he saw increasing bureaucratization as unavoidable in industrial society and recognized its progressive nature as well as its negative consequences. Liberals tend to celebrate 'the miracle of the market' and cast a blind eye towards the bureaucratic nature of enterprises. Not surprisingly, they see industrialism in overwhelmingly positive terms. Weber's ambivalence towards rationalization derived from a more acute perception of the nature of industrialism. Yet he could, without contradiction, be both critical of liberal positive social theory and its pre-social individuals, and sympathetic to its moral and political concerns (Holton and Turner, 1989, p. 20).

Socialism is seen as a *continuation* of the rationalization process, not as a break with it. The more we try to control society the more we are enslaved by its rationalizing processes (Bauman, 1990, 1992). Socialism is therefore a kind of 'iron cage' rather than a source of liberation. This is of the first importance in challenging socialist views, especially those which see nothing inherent in Marxism or the socialist project which might account for the totalitarian character of state socialism. The bureaucratic character is not merely the result of a 'wrong-turning' taken in the development of socialist societies but is a consequence of pursuing rationalization at a social level without regard for catallaxy. Nor can attempts to increase democracy escape the dilemmas, for mass democracy *entails* bureaucracy (Gerth and Mills, 1968, p. 224). Bureaucratization and the individuation of persons have derived not only from processes internal to industry or capital but also from popular struggle for social rights and universalistic standards of social welfare (Abercrombie et al., 1986, pp. 154–5).

Weber's analysis of rationalization and bureaucracy illuminates the nature of both capitalist and non-capitalist organizations, the similarities between these having always been an embarrassment for Marxism. Conflicts between substantive and formal rationality are treated not merely as products of the capitalist character of organizations but as 'transcontextual dilemmas in social life' (Holton and Turner, 1989, p. 54). They take particular forms under capitalism, such as the conflict between the substantive needs of workers and the formal rationality afforded by the existence of wage labour and labour markets in allowing capitalists to select the appropriate workers for particular jobs. But as we indicated in chapter 7, conflicts of substantive and formal rationality are as endemic to socialist societies as to capitalism, albeit in different form. In line with our argument against the exclusion of economizing from positive and normative political economy, Weber pointed to the danger of severing decisions 'from all direct relationships with costs and consequences' and to the need to place checks on the moral imperatives of socialist forms of substantive rationality if formal rational calculation was to take place (Holton and Turner, 1989, p. 46). This could of course be used to legitimize the *status quo*, but a bad use of the argument need not drive out good ones; not all objections on grounds of costs to meeting substantive goals need necessarily be disingenuous.

The similarities between capitalist and non-capitalist organizations are further illuminated by reference to Weber's comments on ownership and control. As we saw in chapter 6, Weber was right to argue that the loss of individual control over activites occurred with the development of all large-scale organizations by virtue of the use of large-scale technology and the need for cooperation and discipline. Our discussions of the problems of 'social ownership' lend *qualified* support to Weber's claim that instead of reversing expropriation, a 'socialistic economy' merely brings to completion the expropriation of individual producers by adding the expropriation of private owners. However, this does not mean that the technical requirements of large-scale production or administration render the social relations of production unimportant; these still affect workers' security, remuneration and control.

A standard criticism of Weber's theory is that bureaucratic control is not as effective as Weber assumed (Lee and Newby, 1983; Holton, 1992). As Giddens (1982, p. 202) points out, bureaucratization does not succeed in centralizing power because its tight chains of dependency and the lack of redundancy in the organizational structure give certain strategically placed groups leverage which can be exerted against those above, whereas less centralized systems, such as markets, may be harder for subordinates to influence. Moreover, the phenomenon of the work-to-rule as a device for *slowing down* work indicates that bureaucratic rules are not and generally cannot be observed to the letter if organizations are to be effective. Instead they need supplementing and sometimes overriding by practices which are more flexible and better adapted to the social relations between workers. While bureaucratic forms of organization are hard to avoid in organizations, they are not sufficient to make them successful. But Weber's overestimation of the power of rationalization and its bureaucratic form is not limited to his analysis of the internal operations of institutions, within their technical divisions of labour, for he also overestimates the tractability of an advanced social division of labour. Like Marx, though less surprisingly, he has an insufficiently materialist understanding of its character and of the barriers to rationalization posed by division of knowledge. To bureaucratize Hayekian economies* is one thing, to do the same for a catallaxy is quite another. As a result, Weber underestimates the anarchic character of industrial, and especially capitalist, societies.

Marxist and Weberian approaches therefore both have visions of progress towards a fully rationalized society. The difference is that while, amazingly, Marx, at least in his later work, and the early Marxists saw no harm in the idea of society as 'one big factory', Weber, more understandably, was clearly deeply disturbed by it. At times, Weber also seems to have shared the idea that such a state was feasible (Weber, 1972, p. 69). On this subject, however, he appears to have wavered. He recognized that rationalization was not limited to the concrete form of bureaucratization in organizations, but also developed through, and depended on, the rise of competitive markets and accounting using market prices. Presumably as a result of his contacts with early Austrian economists,[7] Weber believed that central planning was subject to serious formal irrationality – perhaps to the point of being *in*feasible – since

[7] Concern with the division of knowledge only developed in Austrian economics later.

prices would not be market prices based on competition, and hence the architects of the Plan would not know how to allocate resources (Bellamy, 1992; Weber, 1947, p. 198).[8] While he was therefore aware of the problems of central planning, he never adequately reconciled this with his view of bureaucracy and the iron cage. Even though he believed that formal, calculative rationality depended on market prices, and market prices depended on competition, and competition, in turn, implied multiple centres of power, he did not draw the conclusion that this would limit the extension of bureaucracy and block the realization of the iron-cage scenario. In this, he failed to anticipate Hayek on the impossibility of knowing the ends and means held by individuals (Holton and Turner, 1989).

My argument, and hence my weak or moderate version of the industrialism thesis, is thus quite different from that of Weber and theorists of rationalization, for it emphasizes the intractable and anarchic element of industrial societies. It would be excessively charitable to argue that Weber's theme of the conflict between formal and substantive rationality fully covers the anarchy and uneven development of capitalism, for apart from the heady level of abstraction, these concepts do not take sufficient account of the material qualities of the diverse activities, the associated economies and diseconomies of scope and scale, that have to be rationally coordinated.[9] Weber was too much of a rationalist and not enough of a materialist to grasp this. Although his analysis of the tensions between formal and substantive rationality offers tantalizing glimpses of fundamental dilemmas of modernity, these and other concepts are simply too 'thin' to bear much weight for the purposes of political economy. Many conflicts and problems can be 'glossed' by reference to these tensions, but without necessarily illuminating how the mechanisms actually responsible for them work. Similarly, the identification of particular forms of rationality adds a dimension which is absent in *Capital*, but their embedding in material practices is underplayed; capital accumulation does indeed depend on capital accounting but it is not reducible to it. The taxonomic style of works such as *Economy and Society* yields concepts which may be interesting in themselves but which do not easily connect to one another or to specific contexts. Moreover they give little impression of economic dynamics. The contrast with Marx's method of analysing objects in terms of their internal relations to others and as 'moments' (in the circulation of

---

[8] 'Where a planned economy is radically carried out, it must further accept the inevitable reduction in formal rationality of calculation which would result from the elimination of money and capital accounting.' (Weber, 1947, p. 198).

[9] As Löwith points out, Weber occasionally acknowledged the anarchy of economic production and defended it in terms of the defence of individual rights (Löwith, 1962).

capital), thereby focusing on dynamics, could hardly be stronger. Weber's distaste for general theory prevents him from grasping the connectivity of capitalist processes, while his ideal-type methodology allows him to excuse himself biases, omissions and 'thinness' (Sayer, 1992a, pp. 237–8). For these reasons, Weberianism's contribution to political economy is more as an occasional corrective than as an alternative to Marxism.

Postmodernist–Liberal Affinities

To those accustomed to thinking about social theory in terms of modernity and postmodernity, my concerns with political economy and negotiations between liberal and radical approaches are likely to seem somewhat dated. In some academic circles, radical interests in capitalism have largely been replaced by postmodernist themes such as hostility to grand narratives and modernist rationalization, and the celebration of local knowledges, fragmentation and difference. Those who are committed to the discourse of 'modernity' and 'postmodernity', and those committed to political economy and capitalism tend to regard each other with suspicion. For the former, political economy evades and suppresses issues of difference, new social movements; it's a way of not talking about gender, race and ethnicity, cultural imperialism, and so on. For the political economists, the discourse of modernity and postmodernity is a way of not talking about political economy and the painful questions that lie in the wake of the defeat of the Left, the demise of state socialism, and the rise of the Right. While this book is intended to counter the neglect of abstract political-economic theory, I sympathize with the fears of both points of view, and I see no reason why the two sets of concerns cannot be brought together in more concrete studies. The focus on modernity recognizes features of contemporary society which go beyond capitalism to all industrial societies (e.g. secularization, rationalization, etc.), and counters political economy's silence on culture. However, as I suggested at the outset, the rise of the discourse of modernity in radical academia at a time when the Left was in retreat on the political economy front was surely not innocent. Just when it was becoming more difficult for the Left to continue to attack capitalism, the discourse of modernity offered a more evasive way of talking about capitalism and culture, or industrialism and culture.

As we noted in chapter 1, postmodernism may be dominant in certain parts of academia, but outside academia it is liberal or neo-liberal ideas which have made the running since the late seventies. While there are significant innovations in postmodernist thought, it also has many similarities to liberalism, perhaps as a consequence of their shared political context. These resemblances are rarely noted. One reason for this is that the discourse of postmodernity is overwhelmingly cultural in content, whereas liberalism is political and economic. Another reason is that the discourse of modernity and postmodernity has been strongest in sociology, and historically, liberalism has been sociology's 'other' or foil, albeit a rarely acknowledged one today. Liberalism has a weak social theory, characterized not only by an undersocialized view of action and individuals but by a remarkable lack of concern for the need to test its propositions (Arblaster, 1984), a feature which is even more astonishing in the light of its association with empiricism. It is perhaps because of this historical opposition between sociology and liberalism that some sociologists sympathetic to postmodernism underestimate the liberal, anti-authoritarian strand within modernity and modernism and exaggerate the tendency towards rationalization. For example, Bauman argues that for modernity the world was an 'object of willed action', with sometimes disastrous results; in his view the Holocaust was a consequence of rationalization and state socialism was the most extreme form of rationalization (Bauman, 1992, p. x). But modernity has always embodied elements which counter its rationalizing tendencies – so much so that, while some have worried about the iron cage, others, like Marx and Polanyi, thought that it might disintegrate into anarchy and barbarism. Liberalism's positive social theory may be weak, but this is no excuse for ignoring its crucial role in modernity. Where liberalism has been written out of the script, the resurgence of characteristically liberal themes has had to be interpreted in postmodernist terms. Thus, if we follow Bauman's plausible view that postmodernity is modernity coming to terms with its own limitations, then the postmodernist critique of rationalization can be seen as having been substantially anticipated by liberalism.

In postmodernism we find incredulity towards meta-narratives (Lyotard, 1984), in liberalism hostility towards grand theory (e.g. Popper, 1962). Complementing these themes is a preference for local knowledges in postmodernism and for modest theoretical conjectures and piecemeal social engineering in liberalism.[10] To be sure, the routes to these positions are different – in one case it is the linguistic turn and the association of knowledge with power, in the other, critical rationalism – but the effects are strikingly similar. It must be said that there is an immense irony in the postmodern incredulity towards meta-narra-

---

[10] Although Hayek cites Popper approvingly, Popper is not as anti-constructivist or fatalistic as Hayek. Interestingly, an obituary for Popper (*Guardian*, 20 Sept. 1994) describes him as an 'early postmodernist'!

tives, in that the discourse of modernity and postmodernity is itself a meta-narrative, one which outspans Marxism's grand narrative and is frequently deployed to situate events in the most cavalier manner. The sweeping assertions and portentous, olympian tone of so much talk of modernity and postmodernity is no substitute for analyses of specific mechanisms. Those who are sceptical of meta-narratives should turn their suspicions on 'modernism' and 'postmodernism' first.[11] As Marx observed in the *Poverty of Philosophy*, idealist discourse reduces concrete problems of material actors to expressions of intellectual problems and principles (Marx, 1956).

Foucault's line in attacking totalizing theory is closer to our own, not calling for a rejection of Marxism *per se* but arguing for a de-totalized version (1980, p. 81). Totalizing versions are threatened by references to contingency and catallaxy, the former challenging them at the level of ontology, the latter at the level of substantive theory. At the level of methodology, our strategy of asking counterfactual questions, and examining critical standpoints also has a de-totalizing effect, for it reveals that many of the propositions of Marxist theory of capitalism identify features which are not unique to capitalism. This has the effect of loosening up the theory's tight networks of interdependencies. A similar effect is evident in 'analytical Marxism', which aims to break Marxist theory down into its constituent propositions, so as to evaluate them one by one, though its success in this respect is limited by its casual mode of abstraction.

The similarities between postmodernism and liberalism regarding grand theory extend to common problems. Firstly, their hostility to grand theory reproduces the *a priorism* of its enemy, totalization. Whether we need grand or modest theory for particular explanatory purposes is an *a posteriori* matter. Local knowledges are appropriate for objects which are indeed local, grand theory for objects which are large or widely replicated. It seems likely that the social world consists of both kinds of objects, sometimes linked. It is as absurd and dogmatic to rule in favour of discourses which make it difficult to discover whether there are any grand structures, as it is to privilege theories which refuse to countenance the possibility of fragmentation.

In both liberalism and postmodernism there is a fear of the large-scale

---

[11] Liberals and postmodernists frequently confuse their grounds for objecting to grand theory – sliding from objections to its scale to objections to its allegedly foundationalist character. However, while dogmatism and grand theory often go together, there are rival grand theories and advocates of any one don't have to assume any privileged access to some absolute truth. Grand theory need not be foundationalist. It's also worth noting that postmodernists sometimes grant local knowledges a privileged, foundational status (Sayer, 1993).

political-economic structures presaged by grand narratives/theory, especially the totalitarian society that many have alleged is implied by Marxism. This extends to a hostility towards many other large-scale, technically rational, political and economic projects and organizations, especially those that involve the state. Postmodernist criticism of large-scale, austere, technically rational, modernist architecture and urban planning, and preferences for playful, heterogeneous urban landscapes, resonate with liberal anti-constructivism and anti-planning movements (Harvey, 1989; Thornley, 1990). Liberalism's reluctance to acknowledge the existence of large organizations in its positive theory appears to reflect its suspicion of them in practice, and an apparent longing for a nineteenth-century form of small-scale competitive capitalism. But dislike of large-scale organizations and projects is not all that is at stake; more importantly there is a shared hostility to totalitarianism and repression of difference and dissent:

> post-modern critique of hierarchy, unitary notions of authority, and the bureaucratic imposition of official values has a certain parallel with the principles of toleration of difference in the liberal tradition.
>
> *(Turner, 1990)*

Just as we noted that Hayek's extreme anti-constructivism led to a kind of conservatism or fatalism, so postmodernist rejections of modernist ideas of social improvement can lead to fatalism and quietism.[12] In both cases, but especially in postmodernism, it is a fatalism which is often worn lightly – in keeping with the elite nature of the authors and their audience.

While noting these affinities, we must nevertheless acknowledge that postmodernist thought goes beyond anti-constructivism and its premises – not only criticizing the notion of making society transparent and subject to a single will or collective historical subject, which liberals oppose, but undermining the concept of the unitary, individual, rational subject which has been so central to liberalism itself. I shall have more to say about these tensions and affinities shortly, but I shall do so in the context of another object of liberal and postmodernist scepticism – critical social science.

---

[12] Fay comments:

> it is also an error to think that there is nothing humans can learn about themselves, or nothing they can do to improve their lot. Concentrating solely on the positive side of illumination and activity is naive and leads to an unfounded hubris destined to end in tyranny. Concentrating solely on the negative side of concealment and dependency leads to self-fulfilling despair which insures that the darkness and weakness it fears will in fact prevail. (Fay, 1987, p. 215)

Critical Social Science and Normative Theory

In chapter 2 we argued for greater consideration to be given to alternatives in political economy, both for exploring counterfactuals implied in explanations and for examining the implicit standpoints of critiques. Counterfactual questions help us assess explanations; in particular they help us identify how historically specific social phenomena are. Thus if urban problems are attributed to capitalist social relations of production and the accumulation imperative, then asking what would happen with non-capitalist social relations enables us to consider how far they are indeed specific to capitalism. If Marxism had grasped the mechanisms which perpetuate capitalism, then it ought to have been of use to socialist economists in understanding how to block those mechanisms and replace them with superior ones. The fact that it did not do so, and was actually misleading for that purpose, does indeed reflect badly on its critical theory of capitalism. A critical social science like radical political economy therefore needs to examine alternatives not only as counterfactuals for the purpose of sharpening explanations, but for selecting feasible and desirable critical standpoints so that its critiques carry weight: the two purposes are compelmentary.

Radical social science has traditionally made light of the difficulty of knowing from what standpoint one should critically explain/evaluate social practices. The rise of new radical concerns and the new Right have shaken up the simple, complacent political and ethical beliefs of the traditional Left – and not before time. The decline of the Left in the last decade or so is reflected in the increased uncertainty among radical social scientists regarding both their theoretical allegiances and appropriate political programmes. We have already seen that there are fundamental inconsistencies in the critical standpoints of Marx and Engels' analysis of capitalism, but there are many more inconsistencies in the implicit or explicit critiques made in contemporary social science. Is the problem of industrial decline one of ineffective competitive behaviour, or a consequence of the very existence of a competitive framework? Do we say of a backward region or country that there's a lack of growth, or that growth is the problem? The rise of green critiques of growth are particularly important in upsetting traditional, radical assumptions in favour of industrialization. If industrialization on a par with contemporary rich countries is a positional good at the level of the world (Altvater, 1993), then it must be irrational to call for its extension. One of the hallmarks of the structuralist influence on radical thought was that it taught theorists to see problems as deriving not from the way particular players played the game, but from the very structure of the game itself.

Valuable though this distinction is, there now seems to be less confidence about it, perhaps because, even though the problem might lie with the structure or game, we don't have a convincing alternative game, so we revert to seeking better strategies within the existing one. We saw in the case of rationalization, anarchy and liberty how there were trade-offs between different strategies, such that it was difficult to see how improvements on one front could be realized without creating new problems on another. Often the problems nest within one another and we criticize from the standpoint of an alternative which we think is feasible, even if it worsens a larger problem which is more difficult to resolve. Thus it is tempting to suggest better ways of industrializing, which may be feasible on a limited scale, even though one knows it is worsening the global situation.

It should be abundantly clear that the more or less implicit belief of critical social sciences, such as radical political economy, that contradictions and dilemmas could be successively eliminated without creating new ones, is untenable: it is a modernist myth. There are always going to be trade-offs, though not necessarily zero-sum games, and gloomy though this may sound, we stand more chance of success being aware of this than we do imagining that they don't exist. But there is a further problem with critical social science's confident view of emancipation. This is its assumption that emancipation comes about solely or largely through removal of obstacles – be they illusions held by people which help perpetuate oppressive social practices, relations of domination or material deprivation. Apparently, once we have eliminated these and people can relate to one another freely and as equals, people will be emancipated.

There are several problems with this. Firstly, as we saw in the analysis of markets, good and bad features of social practices may be interdependent rather than separable. Secondly, it is a peculiarly lopsided view of the good society which only considers it in terms of freedom from obstacles and ignores the question of responsibilities, or even renders responsibilities in wholly negative terms as inevitably, rather than contingently, oppressive. In this respect, critical social science is ironically complicit in one of the most fundamental problems of modern society – the concept of emancipation as escape from responsibilities. The more libertarian philosophies, with their celebration of the free, unencumbered, implicitly male individual tend to imply this. Marxism emphasizes and applauds the social individual, but its silence regarding responsibilities and norms, coupled with the popular negative associations of responsibilities as burdens, means that it fails to oppose the notion of emancipation as freedom from responsibility for others. Of course, there are good grounds for the negative associations. Talk of responsibilities *should* arouse suspicion: *whose* responsibilities do we mean? Support for the idea of responsibilities is often associated with conservative discourse, as a covert way of endorsing the currently unequal distribution of responsibilities, especially in relation to gender. But the acceptance of the *concept* of responsibilities does not have to have this conservative subtext, in fact it is a precondition for removing the inequalities relating to responsibilities. Responsibilities can't be eliminated without inducing social breakdown and they won't be borne more equally until they are taken seriously as a subject of moral and political discourse.

Thirdly, as a generalization of this last problem, critical social science gives the impression of the good society as a space cleared of illusions and oppressive relations, in which individuals or groups will naturally find liberation. This implicit view of emancipation is ironically reminiscent of the libertarian concept of negative freedom, i.e. as freedom from interference from others or from the state. But a positive conception of the social good is also needed. Even if the obstacles and relations of domination were removed there are many different forms which an alternative society could take, and there is little incentive for changing from our present society if we have no idea what an alternative society could be like. Moreover, in the event of the removal of existing oppressive relations and practices, specific structures and mechanisms are likely to be needed to prevent the re-emergence of various tyrannies and injustices. The removal of domination, illusions, obstacles and problems is not enough; alternative frameworks are needed. In Habermas's ideal speech situation, social relations are characterized by undistorted communication, power is equalized and the only force is the force of the better argument (Habermas, 1972). Aside from the problem of deciding what constitutes the latter, this still supports the image of the good society as an empty space in which people collectively and freely negotiate a just social order, as if such a situation would be proof against tyranny. The naïvety of this reminds one of the graffiti: 'Blessed are the meek, for they shall inherit the earth – provided that's alright with everyone else.' Instead of addressing the inevitable opacity and cross-purposes of a catallaxy, and the *advantages* of impersonal social coordination via markets, it gives the impression that society – presumably an advanced one – could and should be made transparent and subject to *ex ante* control (Holton and Turner, 1989, p. 6). While Habermas criticizes Marxism's reduction of action to labour and its disregard of communicative interaction, he implicitly endorses its modernist, constructivist project of making the social world a product of design.

Behind these problems of radical political economy is a more general one concerning the relationship between positive and normative theory in critical social science. Without considering normative questions of what ought to be the case we can hardly define what constitutes a problem, and hence how problematic situations are to be explained. Consider the case of childcare. Deciding whether there are any problems here depends on what we think ought to exist as regards childcare and who ought to be responsible for it. This will affect whether we think the problem is one of lack of state provision of nurseries and after-school clubs, etc.; or lack of involvement of *men* in childcare; or a failure of parents to make adequate provision and sacrifices for bringing up their children.[13] Note how closely the definitions and explanations of the problems are bound up with normative judgements about what ought to happen, and where responsibility ought to lie. Not surprisingly, the very selection and definition of a problem implies a value standpoint. We can challenge the explanation of a problem by criticizing the normative standpoints they imply; thus, in the childcare example, the first explanation, with its references to lack of state provision, could be criticized by reference to the implicit standpoints of the second and third.[14] The idea of developing a critical explanation of such situations without examining normative judgements about them is absurd. And yet both in its practice and in the philosophical reconstructions and defences, critical social science is treated as producing evaluations of social situations without recourse to normative theories. Remarkably, it is rare to find more than a mention of ethics in the literature on critical social science (e.g. Fay, 1975, 1987; Bhaskar, 1979, 1989).

The positive (descriptive and explanatory) and the normative (critical and evaluative) sides of critical social science are therefore out of balance; while the positive side is endlessly debated the normative side remains implicit and unexamined. The gulf and the imbalance between the two elements are intolerable. If critical social science is to become more successful it needs to be in closer contact with this normative theory.[15] In the case of Marxism there is historically an explicit aversion to normative theory. Yet we are locked into the particular competitive games of capitalism not simply because we live in an all-enveloping capitalist society, but because desirable and feasible alternatives are so weakly developed. The challenge 'show us a better game' is hard to answer, and it is this inability which most embarrasses the Left

---

[13] Note that this view is not the exclusive preserve of the political Right.

[14] Critical social science is not relieved of normative judgements by taking a 'target group's definition of problems on its own terms, for it is open to critical social scientists to argue that they are mistaken.

[15] Recognition of the importance of normative theory is beginning to develop on the Left. See the important collections by Squires (1993) and Osborne (1991).

and reduces its 'critical' social science towards the striking of critical attitudes. By ironic contrast, liberalism, which is notably weak as a positive social or political-economic theory and is generally considered to be reactionary and apologetic, has a far more developed normative side, as indicated in its sophisticated discussions of the interrelationships between liberty and equality (e.g. Gray, 1986; Kymlicka, 1990; Nozick, 1974; Plant, 1991; Rawls, 1971). Perhaps significantly, the more recent critical social sciences of feminist and green theory suffer less from the gulf between normative and positive, with theorists on either side acknowledging the other (e.g. Barrett and Phillips, 1992; Dryzek, 1987; Dobson, 1990). Nineteenth-century political economists and social theorists had few qualms about combining normative and positive within the same writings. This need not necessarily be considered a problem. There is always a danger of our values regarding what *ought* to be leading us into wishful thinking regarding what *is*, whether we happen to make our normative views available for discussion or not, though if we do, it can make this kind of bias easier to detect.

The normative theory most relevant to political economy is that of political philosophy, itself a branch of moral philosophy. Political philosophy is concerned with defining or interpreting concepts of the political good. What kinds of social practice and organization are good and why? What ought the role of the state to be? Do we have rights and if so what are they? Likewise what are our responsibilities? Typically it explores concepts of the good and works out – at least in idealized terms – what the implications of acting upon those notions would be. For example, Nozick starts from the premise that people have certain property or 'holdings' to which they are entitled, and works out what is likely to happen if they are free to dispose of them as they like, in particular how this might influence the way we evaluate inequality (Nozick, 1974). Political theory is less abstract and deals with actual institutions such as representative democracy or markets, and explores their character, preconditions and consequences. While this might be intended to be primarily positive rather than normative, it cannot help but be of great interest for informing normative judgements of social practices. It is particularly valuable for making us question cherished but unexamined beliefs. Take democracy, recently enjoying renewed popularity on the Left. It is hard to resist the conclusion that democracy is often seen as an unqualified good, such that no reasonable person could question centring political objectives on democratizing society. However, political theory shows that democracy is good and bad; besides virtues it has limitations and disadvantages, and is vulnerable to many abuses (Dunleavy, 1991). These kinds of inquiry help us think about alternatives; they examine at a fundamental level the assumptions that both laypersons and critical social sciences tend to take for granted in defining problems.

However, I realize that this literature is likely to seem utterly alien to those who are accustomed only to 'normal' descriptive and explanatory social science – which I take to include radical political economy. There is therefore almost certain to be considerable resistance to the idea of relating political economy to debates in political philosophy, with its daunting thickets of arguments about liberty, equality, rights, responsibilities, entitlements, etc. The most common objection is to the lack of foundations for normative discourse – why these values, these premises? Why should we accept this as good? The critique of foundationalism applies to normative theory as much as to positive, explanatory theory. Similarly, just as, in epistemology, some people react to the realization that there are no ultimate foundations by flipping over into judgemental relativism (or 'silly relativism' as Rorty calls it), in which no theory is better than any other, so in response to the lack of foundations for normative analyses, some flip into emotivism or subjectivism, where values are beyond the scope of reason and nothing more than what we happen to like or dislike. However, typically, even those who are drawn towards emotivism continue to *argue* or reason about values in practice. The situation in normative theory is consequently not so different from that in positive science. In neither case does the lack of foundations prevent us arguing about what is better or worse.

Two positions dominate contemporary debate in political philosophy. 'Ethical naturalists' argue that what is good for people depends on their nature. To say that something is good or bad for people is to imply something about what they are like, about their needs. 'Nature' here need not exclude their social nature, nor need it deny that people are transcendent beings able to remake themselves to a certain extent. Nor, furthermore, need it imply epistemological foundationalism, for we need not pretend that we have any privileged access to some absolute truth; claims about our social natures are always going to be open to revision, just as claims about chemistry are always revisable. By contrast, communitarians appeal not to human nature, however qualified, but to the values, conventions and practices of the community in which actors live. Such a view emphasizes that systems of values are socially constructed, but while this is obviously true we cannot be or do or define as good just anything, according to our whims. There are limits to what we can construct ourselves as. Although the disagreements between ethical naturalism and communitarianism are fundamental, neither endorses a subjectivist or emotivist refusal of attempts to reason about values.

A more populist line of objection to normative theory takes the view that we already *know* what we want, and that we know who the enemy is. But even if the working class didn't need Marx to tell them they were exploited, or even if women don't need feminism to know they are oppressed, it is absurd to suppose that lay understandings are as good as Marxism and feminism at explaining the how and why of capitalism and patriarchy. Nothing is more reminiscent of the old empiricist refusal of theory in positive social science – 'we don't need theory, we can just observe'. The answer to those who refuse normative theory is the same as to those who reject positive theory: we need to consider theory because it problematizes lay understandings and offers a more sustained, systematic and critical reflection on the issues. The refusal of normative thinking is a recipe for the failure of emancipatory projects, or worse, a recipe for tyranny.

It may also be tempting to try to resist normative issues by 'sociologizing' value questions out of existence, arguing that values derive from historically specific social arrangements and are nothing more than conventions. There is of course a specifically Marxist variant of this, in which values derive from, and rationalize, ruling interests. Values are indeed intimately associated with particular social practices, though some have a more universal appeal, as is evident in certain transhistorical/transcultural sentiments; the values of others distant in time and space are not necessarily a mystery to us. But in any case, the social embedding of values doesn't absolve us from making judgements about what is good or bad. Without such judgements we wouldn't have a problem to address.

The consequences of these refusals of normative theory are often evident in radical politics. It is common for radicals to say that they are against 'domination'. But as we have seen, domination is not the only source of suffering and inequalities. Many problems are unintended consequences of other people's actions – including people who do not dominate us in the way a master dominates a slave or a capitalist a worker. It has also become common for radicals to say that they are for 'empowerment' of particular oppressed groups. However, in an interdependent world, empowering one group often means losses for another group, and not necessarily only those who dominate them. The question of why anyone should pay for another group's empowerment needs an answer. Rights imply responsibilities. For whom and for what should we be responsible? Why should others be responsible for us and subsidize us? Talk of empowerment can become a way of avoiding these questions. The danger of allying a concern with difference to a concern with empowerment or rights and stopping there is that it invites a new kind of Hobbesian order – of difference against difference. And if this in turn is allied to an emotivist view of values, then we have a new version of 'might – or will – is right'. Without a framework of general principles and rules about rights and responsibilities, then again, the field is open for tyranny. This sounds distinctly modernist and the development of hybridity and cultural pluralism might seem to render agreement about such a framework more difficult, but on the other hand they also make the need for a framework more important.

Finally, there is a different kind of standard criticism of this kind of normative political philosophy, which is that it is too abstracted from the material conditions and social relations in which we live. I shall discuss this later.

The Clash of Political Philosophies

In order to indicate the kinds of issues which need to be addressed in thinking about critical standpoints and alternatives in political economy, I will provide a sketch of the debates between four broad positions – liberalism, Marxism, communitarianism and theories of difference. By theories of difference I mean postmodernist, feminist, post-colonial, anti-racist and new social movement concerns with difference. This group is particularly heterogeneous. I recognize that some feminists are anti-postmodernist (e.g. Skeggs, 1991) or at least sceptical of it (e.g. Hartsock, 1990; Soper, 1991). But what they have in common is a suspicion of both collectivism, associated with Marxism, and individualism, associated with liberalism, on the grounds that both allegedly have the effect of denying or dismissing difference and concealing or perpetuating relations of domination other than those of class. Because these positions are internally heterogeneous the account will inevitably oversimplify the debates. But my main aim is to show that while there are many important mutual criticisms of each of the four positions, something of each of them survives; indeed, each position often presupposes something of the doctrine it seeks to overthrow. The positions have both normative and positive content too; they can therefore be criticized in terms of both sides.

The critique of the liberal individual

One of the most criticized aspects of liberalism is its concept of the individual:

1 Often the individual is treated as pre-social or asocial, a bundle of desires and preferences confronting a society reducible to other such individuals.
2 Liberals typically defend the rights of the propertied individual, not consid-

ering the propertyless whose liberty is restricted by the property of others (Kymlicka, 1990; Haworth, 1994).

3   The liberal individual is actually a man, unconstrained by relationships of dependency, acting in the public sphere.

All of these have been heavily criticized.

Term (1) is of course an absurd characterization of individuals, simply failing to recognize their socialization – their dependence on shared meanings, on a common language and on living in a web social relations. But while the liberal view of the individual is characteristically undersocialized, there is a danger of putting an oversocialized view of individuals in its place in which they are nothing but the intersection and product of those relations and have no power to change them and hence change their identity (Kymlicka, 1989, 1992; Soper, 1991; Wrong, 1961). Our ability to reflect on the relations which constitute us suggests we are not reducible to their intersection (Soper, 1991). To be an agent is not to be able to escape from social relations but to be able to act so as to change *some* social relations by working with and through others. Divorce is an obvious example. The divorcee has broken a social relationship, but has done so by working through other social relations and practices. Divorcees remain social beings, only living in a partly changed set of social relations. This socialized conception of agency is compatible with liberalism's defence of individual liberty, for the latter involves not the retreat from all social relations but the possibility of changing them from within. As some liberals have emphasized, the independent citizen is not asocial but depends on a collection of elaborate social arrangements designed to provide certain kinds of security (James, 1992).

Term (2) is more easily dropped. Doing so allows left liberals to attack conventional liberal views of individual liberty and private property.[16]

Much has been written on (3), particularly in connection with the closely related critique of the public/private distinction (e.g. Pateman, 1987; Phillips, 1991b). Liberalism does not adequately address the situation of women, typically differently placed to men with respect to divisions between private and public, and having less scope for self-regarding actions as opposed to other-regarding actions, by being encumbered by caring relations of dependence.[17] Although liberals don't

---

[16]   While many resort to a concept of positive liberty to do this, as Haworth argues, this is not necessary; one can accept the libertarian concept of negative freedom (i.e. freedom from interference from others) and argue that private property can obstruct the freedom of the unpropertied to act (Haworth, 1994).

[17]   The liberal distinction between self-regarding and other-regarding actions is a distinctly fuzzy one, but as with the fuzzy distinction between the front of someone's head and the back, there are many cases when we can easily make the distinction.

have to assume the individual to be pre-social, they tend to conceive of social relations in terms of interactions between individuals and their 'generalized others'. Our others are entitled to expect and to assume from us what we can expect and assume from them. They are our formal equals and we abstract from differences and concrete individuality. This typifies the public sphere of independent, somewhat aloof individuals, able freely to enter into contracts, and entitled to do what they like provided they do not harm others. It implies a distinctly masculine image of society, one that fails to address relationships between 'concrete others'. The latter are more familiar in the private sphere and in the lives of women, and involve relating to *concrete* others with particular individualities, biographies and identities, assuming obligations and responsibilities that recognize and respond to differences in needs, talents and capacities (Benhabib, 1987; James, 1992). Yet in modern societies we are dependent on vast numbers of people, most of whom are strangers to us, and whom we cannot treat as concrete others. Nevertheless, both kinds of relation are needed.[18] There are obvious resonances between this generalized/concrete distinction and the traditional sociological distinction between *gesellschaft* and *gemeinschaft*. Our relationships in public politics and across the social division of labour, whether coordinated primarily by markets or planning or democratic control, are primarily with generalized others. Relationships with concrete others dominate in families and friendships, and in personal politics. There is often an uneasy combination of the two types of relationship in organizations and networks.

Whatever the limitations of the concept of the generalized other, it seems impossible or dangerous to reject any concept of the individual as a person who, though a social being, has or should have some autonomy (Soper, 1991). It is also important that relations with concrete others can be negotiated and chosen rather than imposed. Thus one of the strongest critics of the liberal concept of the individual nevertheless argues that:

> neither liberalism nor feminism is conceivable without some conception of individuals as free and equal beings, emancipated from the ascribed, hierarchical bonds of traditional society.
>
> *(Pateman, 1987, p. 103)*

Important though these criticisms of liberalism are, then, it is possible both to respond to them and retain something of the key feature of

---

[18] See the excellent article by Susan James (1992) on the links between public and private.

liberal doctrine, namely the defence of the individual as an agent against domination and interference by others or the state.

On the related question of *difference*, we find both resonances and oppositions between liberalism and theories of difference. Liberalism and Marxism expected ascriptive, cultural difference to whither away, in the former case in favour of the independent individual, in the latter in favour of the social individual or collective worker.

> Both liberalism and Marxism have their roots in an 'enlightened' universalism which looked forward to the eventual emergence of a cosmopolitan world society; a global community in which transnational social bonds and universally held notions of peace, justice, equality and freedom defined the conditions of human existence.
>
> *(McGrew, 1993, p. 77)*

This universalism is prefigured in the use of universal, undifferentiated categories like 'class' or 'individual' in the 'modernist meta-narratives' of Marxism and liberalism respectively. Postmodernist theory reacts against the suppression of difference in these theories and attempts to deconstruct categories like 'race' and 'woman'. But while Marxism and liberalism both expected a triumph of reason over difference, liberalism is far less troubled by the persistence and continued evolution of difference. Liberals have often championed individual eccentricity (e.g. Mill, 1859) and although this involves a self-selected difference rather than a socially ascribed difference, as long as the latter does not take repressive forms, it might be accommodated under the liberal defence of the individual.

The celebration of fragmentation and difference in postmodernism resonates with the recognition of epistemological relativism in philosophy and substantive developments such as the rise of multicultural societies and 'hybridization' (Walby, 1994). But there are usually new sources of interdependence as well as fragmentation. The deepening and widening of the international division of labour is itself such a case, so that those whose lives were formerly separate from ours are now linked to our own. There are therefore unifying processes, linking the fate of distant and different strangers. Moreover, the globalization of externalities means that 'late modernity produces a situation in which humankind in some respects becomes a "we", facing problems and opportunities where there are no others' (Giddens, 1991, p. 27).

The liberal principle that all individuals have an equal right to pursue their own interests without interference provided they don't harm others presupposes that all individuals are of equal worth. And yet, liberals are notorious for legitimizing the inequalities which arise in the liberal ideal state of maximum individual liberty. Liberalism's ambiguous stances towards equality have now been complicated by the problem of difference. Whereas inequality might be seen as something which could, and in the view of socialists, should largely be removed, difference can concern characteristics which are positive, rather than negative, such as differences of ethnicity and sexuality. The liberal treatment of individuals as equal ('generalized others') fails to take account of difference. Giving equal treatment to individuals without regard for how their values, identity or biological characteristics differ (for example, ignoring the fact that men can't get pregnant and women can) can make them more unequal. All too often the treatment is only equal in the sense of imposing the norms of the dominant group uniformly on others (though a consistent liberalism would seek a framework which did not impose one groups' norms on others). But then differential treatment, for example, bans on night-working for women, can also reinforce inequality. The discussions within feminism on this, critically evaluating the concept of equality, are salutary (e.g. Bock and James, 1992; McDowell, 1991; Phillips, 1987, 1991a and b; Pateman, 1987; Soper, 1991). Differences in identity, in the extent of involvement of relationships of dependency, such as parenthood, and differences in values and in biology – all make simple appeals to equality dangerously naïve: on the other hand, forgetting equality for difference is no less dangerous. The postmodern concern with difference and marginalized people, and the attempt to make the margins the centre, could be seen as a challenge to the liberal assumption of equality. Yet there is of course no reason for anyone to take any notice of marginalized groups unless they believe that all individuals are of equal worth, and the marginalized have no grounds for making claims against the powerful if they do not also assume this (see Soper, 1991). In this way, the liberal principle comes back on the rebound against the critics of liberalism.

As we saw earlier, socialist collectivism assumes the possibility of a universal collective subject of history, able to control its circumstances through the supercession of class (Laclau and Mouffe, 1985). It therefore demotes other, non-class sources of division, such as gender, race, sexuality, plus the division which theories of difference themselves generally ignore – the social division of labour. Theories of difference draw attention to the way in which Marxism treats the special interests of men as the general interests of the working classes (See Hart, 1989 for a particularly lively critique). Just as Marx attacked the ruling classes for presenting their special interests as the general interests of all, so Marxism – and liberalism, in a different way – can be attacked for presenting the interests of men as the interests of women too. In turn, black women have challenged the right of white women to speak for them, and so on. The universal collective subject has been exposed as a repressive fiction. But then central to liberalism is the doctrine that individuals are the best judges of their own interests, and therefore that others should not presume to speak for them.[19] For liberals, even if the state can demonstrably run things better than individuals can, it should not be automatically assumed that it should be allowed to do so, for paternalism is in general wrong, even if it occasionally does good (Mill, 1859). People should be allowed to determine their own futures including being allowed to make their own mistakes. This is a common demand of teenagers of their parents, and cannot easily be dismissed.

Communitarianism rejects universal concepts of individuals, human nature and human needs, and argues that needs and conceptions of rights and obligations are relative to the particular communities in which they are socially embedded; external judgements are therefore disqualified. This might initially seem an attractive doctrine in the face of value pluralism and in response to post-colonial critiques of ethnocentrism. However, while values and needs are strongly influenced by social context, there are many problems with communitarianism. It is hard to identify exactly what a 'community' is supposed to be in the context of a world economy of multi-ethnic societies, in which people act in a variety of different settings. Communitarians are notorious for evading requests for examples of their key concept.[20] But while communitarianism accepts that there are fundamental differences between communities it cannot easily admit that different communities are interdependent and overlap, and are internally heterogeneous or characterized by difference, thus giving scope for dissensus over the values according to which actions are to be judged.

Far from being a progressive or permissive stance, allowing cultural diversity, it can become a conservative doctrine, resistant to the presence of dissensus and power within communities and reluctant to question their dominant values. Typically the supposed common values in a community are merely the values acceptable to the dominant group (Kymlicka, 1990). Communitarianism both relativizes and neutralizes ethical questions, reducing them to matters of 'what we do around here'. The only court of appeal is to the conventions and norms of the community, and it is hard to see how the norms themselves could be challenged. If a group within the community claim they are oppressed

---

[19] There are surely occasions when our others cannot make themselves heard and hence when it is irresponsible not to speak for them.

[20] Rorty refers to groups of miners and philosophers as communities, though he acknowledges that we move in many different communities and that this is the main source of moral dilemmas (1983).

and call upon their oppressors to desist, they can be dismissed with the answer that the situation is acceptable according to the conventions of the community. In response, communitarians may be tempted to redefine community so as to externalize dissent, but then if we follow the logic of that strategy we end up by treating those who are locked into conflictual interdependent relations as if they existed in separate and improbably small communities, thereby making the grounds for the antagonisms unintelligible. In politics, precisely what 'our' community is and who is in it, and who outside, is itself contested. Furthermore, if our values are defined by our social context it is difficult to see how we could find them unacceptable. Unless there are human needs which are not totally defined by community it is hard to understand both why there should ever be internal dissent and why there are so many similarities between the value systems of different communities.[21] One possible explanation of the latter is that much social action does not involve communities in the more concrete sense. Neither the implicit abstract notions of community in philosophy nor those referred to in actual politics take account of the situation of most people within a global division of labour in which strangers depend on one another. Here, however, communitarians could exploit the elasticity of their central concept and argue that liberal society is itself a community in terms of having shared values, in particular those associated with the 'cult of the individual', in turn a product of an advanced division of labour and Durkheimian 'organic solidarity' (Bellah, 1973; Rorty, 1983). Liberal critiques of communitarianism strike deep but some kind of broad, communitarian context cannot be eliminated altogether, for liberalism itself presupposes one.

Nevertheless liberalism is resolutely opposed to anything more than a minimal communitarian element – one that provides procedural rather than substantive norms, so that individuals are free to follow their own conception of the good, provided they do not harm others. Liberal suspicions of community have much in common with their opposition to totalitarianism and their warnings about the tyranny of the majority under democratic rule: communities can be paternalistic and repressive.[22]

While many have noted the affective, honorific power of 'community', few have adopted as critical a stance towards it as Iris Marion Young (1990). Young's deconstruction of the concept emphasizes the exclusive character of community as a form of social closure, a way of distinguishing one's own group from others. The flip side of community is xenophobia and racism.[23] Male violence against women exposes as myth

[21] In response to this some communitarians are willing to admit a few universals, e.g. Winch, 1978.
[22] See, for example, Fukutake's account of repressive communitarianism in Japan (1982).
[23] I am grateful to Mike Keith for recommending Young's work.

the idea that the unmediated, direct, face-to-face relationships valued by advocates of community are necessarily benign. Conversely, mediated, distant relationships are not necessarily problematic. Furthermore, communitarianism implies that respect for others must be based on empathy, mutual understanding and shared norms. This simply does not address the situation of complex, multicultural societies where we have to coexist with and respect strangers without the benefit of empathy and shared values (Young, 1990; Kymlicka, 1989, 1992). The alternative vision of an urban society put forward by Young incorporates 'openness to unassimilated otherness' (p. 319); 'persons live together in relations of mediation among strangers with whom they are not in community' (p. 303). In such a vision, relationships with concrete others are comparatively limited in scope to households and friendships. This invites comparison with the views of economic liberals and their favourite story of the miracle of the market coordinating the actions of strangers, reconciling the pursuit of self-interest of each with the interests of others. According to Plant (1991), neo-liberals have actually played the multicultural card in criticizing the communitarian character of socialism's end-state values. Unfortunately the reality of markets, especially when coupled with capitalist social relations of production, is that they also create, exacerbate and take advantage of inequalities and relations of domination, reproducing divisions and uneven development. Unless regulated they have a corrosive effect on social solidarity. Thus despite liberal intentions they provide fertile soil for racism.

In the absence of any consensus regarding human nature and needs or on whether it is even sensible to consider such questions, a liberal view of politics is bound to flourish, for it assumes that there can be no agreement on the nature of the human good. Hence it takes 'the aim of political organization and authority to be not so much the realization in the sphere of politics of one particularly contested conception of human excellence, but rather the securing of a framework within which human beings can pursue their own self-chosen view of the good, whatever it may be, so long as it does not conflict with the pursuit by others of their good.' (Plant, 1991, p. 67). However, liberalism's problem has always been that the individualistic pursuit of private goals does not necessarily produce desirable results at a social level – outcomes could be seriously dysfunctional for many.

Whether we are hopeful or not about the prospect of societies with fundamental value divisions, whether it is realistic to see them as potentially harmonious or as bleak settings in which people live at best in resigned and fearful mutual tolerance or at worst in inter-ethnic conflict, one thing is clear – the cosy vision of a homogenous community of interests is irrelevant for much of modern life, no matter how well-meaning its proponents and no matter how comforting the vision may be. (Even Young concedes its appeal.) And it's largely irrelevant not only because of the multi-ethnic nature of modern society but because an advanced division of labour, stretched out around the world, supports not communal relationships but relationships between strangers. Community may exist within and support a Hayekian economy* but a catallaxy cannot be a community in any strong sense.[24]

We noted earlier that political philosophies, such as those of communitarianism or liberalism, have long been criticized for losing contact with the empirical analysis of actual social organization. Thus, liberalism has traditionally abstracted from caring relationships, from power and dependency and from the social constitution of actions in general, presenting an image of society as an aggregate of individuals pursuing their private notions of the good, enclosed within their private property, on which others may not encroach. Though such an image might superficially appear to be objectified – for example, in the suburban environment – our actions are so tangled up with those of others, and the ways in which they affect and constrain others so complex and indirect, that it is extraordinarily difficult to decide whether individuals' actions harm others. Liberalism's social theory is thin and ideological – too concerned with idealizing markets and minimizing state action to want to find out what happens in existing societies. But this is not the only example of the sociological naïvety of political philosophy, induced by its lack of contact with concrete social science; as Young notes, arguments about community are no less abstracted from the fabric of economic relationships present in modern, urban society.

However, no matter how frustrating we find this loose or excessive abstraction in political philosophy, we cannot simply drop it for more empirical studies of actual organizations. As Held puts it,

> if a theory of the most desirable form of democracy [or any other form of social organization] is to be at all plausible, it must be concerned with both theoretical and practical issues, with philosophical as well as organizational and institutional questions. Without this double focus, an arbitrary choice of principles, and seemingly endless abstract debates about them, are encouraged. A consideration of principles, without an examination of the conditions for their realization, may preserve a sense of virtue, but it will leave the actual meaning of such principles barely spelt out at

---

[24] In Durkheim's terms, occupational groups could approximate a community, countering the effects of anomic and forced forms of the (social) division of labour (Durkheim, 1964). Although Durkheim was concerned to situate morality in particular social circumstances, even such formulations as these are too lacking in political-economic content to be of much use.

all. A consideration of social institutions and political arrangements without reflection upon the proper principles of their ordering might, by contrast, lead to an understanding of their functioning, but it will barely help us to come to a judgement as to their appropriateness and desirability.

(Held, 1994, p. 308)

Thus if we were to select more specific topics, like schooling or the care of the elderly, we would still not escape fundamental questions like who should be expected to care for the elderly, or whether care should be commodified. The challenge that lies before us of finding feasible and desirable alternatives in political economy requires this double focus. Here Michael Walzer's *Spheres of Justice* offers a kind of political philosophy that is more attuned to the different spheres in which action takes place, and the conflicts that result between them. He is therefore able to deal with moral issues involved in more concrete matters such as immigration and segregated housing. While we cannot expect theory to anticipate the specificities of particular political conjunctures, a normative theory which was more attuned to the patterning of social life could usefully inform political practice, and counter the loss of direction associated with the decline of the Left.

Conclusion

While our critique has far-reaching implications for political economy, it bears noting that it is nevertheless not only abstract but incomplete, in that there are other areas – such as household economies and green political economy – where much more work is needed (but see, for example, Fine, 1992; Altvater, 1993; and Dryzek, 1987). Explanations of concrete situations are likely to have to draw upon theory in these areas too. However, I hope to have shown the dangers in socialist thought of the apparently innocuous abstraction from the materiality and complexity of advanced economies and their divisions of labour and knowledge. In sum, it can lead to a systematic misunderstanding of existing economic systems, their distributions of power and characteristic dynamics and features. By extension, it lends support to alternatives which are infeasible or undesirable. Appreciating the double nature of capitalist and socialist industrial societies does not reveal a clear path to emancipation, for almost certainly there is none; rather it heightens our awareness of the many dilemmas of their development. At the same time, a disaggregative approach, asking counterfactual questions and comparing the attributes of capitalist, market-socialist and state-socialist economies can enrich understanding of what could be as well as what is.

Given the dilemmatic quality of development it will be no surprise that I do not have ready-made prescriptions. What I have suggested instead is some of the sources we can draw upon in reformulating radical political economy. One of these is liberalism, though I am certainly not advocating its adoption in place of Marxism. Rather the point was that liberalism deserves to be given more serious attention than it has recently received in radical academia, firstly because it continues to pose the most direct challenge to Marxism's traditional heartland – the political economy of capitalism; secondly because of its fertile tensions and affinities with feminist, anti-racist and postmodern critiques of Marxism and communitarianism; and thirdly because, as we have seen, its opponents cannot escape entirely from its premises. The discourses of political philosophy – Marxism, liberalism, feminism, communitarianism – are not separate and pristine but deeply intertwined, each one gaining its identity not in isolation but from its relationship to the others. There is much to be gained from exploring the links as well as the tensions between these positions for they are just about the only raw materials we have for developing critical standpoints. A reformulated radical political economy would need to be post-Marxist and post-liberal, but it would need to retain as well as go beyond some elements of the original positions.

Since critical social science has proved to be a much more problematic enterprise than realized twenty years ago (Fay, 1987), it makes sense to examine its standpoints and implicit alternatives more closely – particularly their constructivist and collectivist assumptions. Radical political economy cannot continue to follow Marxism in standing apart from the debates of normative political theory, nor can it embrace a postmodernist celebration of fragmentation and rejection of the search for a better social framework. It is now more clear than ever that struggles which are directed against domination and oppression but which lack any normative direction in terms of alternative frameworks are unlikely to be successful. Though it has recently suffered from neglect we need a radical political economy more than ever before.

References

Abercrombie, N., Hill, S. and Turner, B. S. (1986): *Sovereign Individuals of Capitalism*, London, Allen and Unwin.

Aganbegyan, A. (1988): *The Challenge: Economics of Perestroika*, London, Hutchinson.

Aglietta, M. (1979): *A Theory of Capitalist Regulation*, London, New Left Books.

Allen, J. and McDowell, L. (1989): *Landlords and Property* Cambridge, Cambridge University Press.

Altvater, E. (1993): *The Future of the Market*, London, Verso.

Ambrose, P. (1986): *Whatever Happened to Planning?*, London, Methuen.

Arblaster, A. (1984): *The Rise and Decline of Western Liberalism*, Oxford, Blackwell.

Arnold, N. S. (1989): 'Marx, central planning and utopian socialism', *Social Philosophy and Policy*, 6 (2), 160–99.

Arnold, N. S. (1990): *Marx's Radical Critique of Society*, Oxford, Oxford University Press.

Arrow, K. (1974): *The Limits of Organisation*, New York, Norton.

Auerbach, P. (1988): *Competition*, Oxford, Blackwell.

Auerbach, P., Desai, M. and Shamsavari, A. (1988): 'The transition from actually existing capitalism', *New Left Review*, 170, 61–79.

Badham, R. (1986): *Theories of Industrial Society*, London, Croom Helm.

Bagguley, P., Mark-Lawson, J., Shapiro, D., Urry, J., Walby, S. and Warde, A. (1990): *Restructuring: Place, Class and Gender*, London, Sage.

Bahro, R. (1978): *The Alternative in Eastern Europe* London, New Left Books.

Baland, J.-M. and Platteau, J.-P. (1993): 'Are economists concerned with power?', *IDS Bulletin*, 24 (3), 12–19.

Bardhan, P. K. (1992): 'Market socialism: a case for rejuvenation', *Journal of Economic Perspectives*, 6 (3), 101–16.

Bardhan, P. and Roemer, J. E. (1992): 'Market socialism: a case for rejuvenation', *Journal of Economic Perspectives*, 6 (3), 101–16.

Barratt-Brown, M. (1970): *What Economics is About*, London, Weidenfeld.

Barrett, M. and Phillips, A. (eds) (1992): *Destabilizing Theory*, Cambridge, Polity.

Barry, N. P. (1979): *Hayek's Social and Economic Philosophy*, London, Macmillan.

Bauman, Z. (1982): *Memories of Class*, London, Routledge.

Bauman, Z. (1990): *Thinking Sociologically*, Oxford, Blackwell.

Bauman, Z. (1992): *Intimations of Postmodernity*, London, Routledge.

Bell, D. (1974): *The Coming of Post-Industrial Society* London, Heinemann.

Bellah, R. N. (ed.) (1973): *Emile Durkheim on Morality and Society*, Chicago, University of Chicago Press.

Bellamy, R. (1992) *Liberalism and Modern Society*, Cambridge, Polity.

Benhabib, S. (1987): 'The generalized other and the concrete other', in E. Frazer, J. Hornsby and S. Lovibond (eds) *Ethics: A Feminist Reader*, Oxford, Blackwell, 267–300.

Berry, B. J. L. (1967): *Market Centres and Retail Location*, Englewood Cliffs, N.J., Prentice-Hall.

Best, M. (1990): *The New Competition*, Cambridge, Polity.

Bhaskar, R. (1975): *A Realist Theory of Science*, Leeds, Leeds Books.

Bhaskar, R. (1979): *The Possibility of Naturalism*, London, Verso.

Bhaskar, R. (1989): *Reclaiming Reality*, London, Verso.

Billig, M., Condor, S., Edwards, D., Cane, M., Middleton, D. and Radley, A. (1988): *Ideological Dilemmas*, Beverley Hills, Sage.

Blackburn, R. (ed.) (1991a): *After the Fall: The Failure of Communism and the Future of Socialism*, London, Verso.

Blackburn, R. (1991b): 'Fin de siècle: socialism after the crash', *New Left Review*, 185, 5–67.

Bluestone, B. and Harrison, B. (1982): *The Deindustrialization of America*, New York, Basic Books.

Bobbio, N. (1987): *The Future of Democracy*, Cambridge, Polity.

Bock, G. and James, S. (eds) (1992): *Beyond Equality and Difference*, London, Routledge.

Bottomore, T. (1985): *Theories of Modern Capitalism*, London, Allen and Unwin.

Bottomore, T. (1990): *The Socialist Economy: Theory and Practice*, Hassocks, Harvester.

Bourdieu, P. (1984): *Distinction*, London, Routledge and Kegan Paul.

Bowles, S. and Gintis, H. (1986): *Democracy and Capitalism*, New York, Basic Books.

Boyer R. (1990): *The Regulation School: A Critical Introduction*, New York, Columbia University Press.

Braverman, H. (1974): *Labour and Monopoly Capital*, New York, Monthly Review Press.

Brenner, R. (1986): 'The social basis of economic development', in Roemer, J. (ed.) *Analytical Marxism*, Cambridge, Cambridge University Press, 25–53.

Brown, D. and Harrison, M. J. (1978): *A Sociology of Industrialization*, London, Macmillan.

Brus, W. (1972): *The Market in a Socialist Economy*, London, Routledge.

Brus, W. and Laski, K. (1989): *From Marx to the Market: Socialism in Search of an Economic System*, Oxford, Clarendon Press.

Buchanan, A. (1982): *Marx and Justice: The Radical Critique of Liberalism*, London, Rowman and Allanheld.

Buchanan, A. (1985): *Ethics, Efficiency and the Market*, London, Clarendon Press.

Buchanan, J. M. and Vanberg, V. J. (1991): 'The market as a creative process', *Economics and Philosophy*, 7, 167–86.

Burawoy, M. (1979): *Manufacturing Consent*, Chicago, University of Chicago Press.

Burawoy, M. and Kritov, P. (1993): 'The economic basis of Russia's political crisis', *New Left Review*, 198, 49–71.

Burawoy, M. and Lukacs, J. (1985): 'Mythologies of work: a comparison of firms in state socialism and advanced capitalism', *American Sociological Review*, 50, 723–37.

Callinicos, A. (1991): *The Revenge of History*, Cambridge, Polity.

Carling, A. (1986): 'Rational choice Marxism', *New Left Review*, 160, 24–62.

Carling, A. (1992): *Social Division*, London, Verso.

Castells, M. (1976): *The Urban Question*, London, Edward Arnold.

Castells, M. (1978): *City, Class and Power* London, Macmillan.

Castells, M. (1983): *The City and the Grassroots*, London, Edward Arnold.

Cawson, A., Morgan, K., Webber, D., Holmes, P. and Stevens, A. (1990): *Hostile Brothers: Competition and Closure in the European Electronics Industry*, Oxford, Clarendon Press.

Centre for Local Economic Studies (1992): Untitled brochure, CLES, Alberton House, St Mary's Parsonage, Manchester, UK.

Chambers, R. (1983): *Rural Development: Putting the Last First*, London, Longman.

Chandler, A. (1977): *The Visible Hand*, Cambridge, Harvard University Press.

Clarke, S. (1982): *Marx, Marginalism, and Modern Sociology*, London, Macmillan.

Cockburn, C. (1983): *Brothers*, London, Pluto.

Cohen, G. A. (1978): *Karl Marx's Theory of History: A Defence*, Oxford, Clarendon Press.

Cohen, G. A. (1988): *History, Labour and Freedom: Themes from Marx*, Oxford, Clarendon.

Cohen, G. A. (1991): 'The future of a disillusion', *New Left Review*, 190, 5–20.

Collier, A. (1994a): 'Value, rationality and the environment', *Radical Philosophy*, 66, 3–10.

Collier, A. (1994b): *Critical Realism*, London, Verso.

Cooke, P. (1989): *Localities*, London, Unwin Hyman.

Corbridge, S. (1986): *Capitalist World Development*, London, Macmillan.

Corbridge, S. (1993): 'Marxisms, modernities and moralities: development praxis and the claims of distant strangers', *Environment and Planning D: Society and Space*, 11 (4), 449–72.

Counter Information Services (1973): *The Recurrent Crisis of London*, London, CIS.

Crouch, C. and Marquand, D. (1993): *Ethics and Markets*, Oxford, Blackwell.

Dahrendorf, R. (1967): *Class and Class Conflict in Industrial Society*, Palo Alto, Calif., Stanford University Press.

Davis, M. (1991): *City of Quartz*, London, Verso.

Delphy, C. and Leonard, D. (1992): *Family Exploitation*, Cambridge, Polity.

Devine, P. (1988): *Democracy and Economic Planning*, Cambridge, Polity.

Dobb, M. (1937): 'The trend of modern economics', reprinted in Hunt, E. K. and Schwartz, J. (eds) (1972) *A Critique of Economic Theory* Harmondsworth, Penguin.

Dobson, A. (1990): *Green Political Theory*, London, Allen and Unwin.

Doel, C. (1994): 'Supply chain governance, the state and the regulatory process', mimeo, Department of Geography, University of Cambridge.

Dore, R. P. (1983): 'Goodwill and the spirit of market capitalism', *The British Journal of Sociology*, 34, 459–82.

Dore, R. P. (1987): *Taking Japan Seriously*, London, Athlone Press.

Dryzek, J. S. (1987): *Political Ecology*, Oxford, Blackwell.

Duncan, G. (1973): *Marx and Mill*, Cambridge, Cambridge University Press.

Dunleavy, P. (1991): *Democracy, Bureaucracy and Public Choice*, Brighton, Harvester Wheatsheaf.

Dunn, J. (1979): *Western Political Theory in the face of the Future*, Cambridge, Cambridge University Press.

Dunn, J. (1984): *The Politics of Socialism*, Cambridge, Cambridge University Press.

Durkheim, E. (1953) [1924]: *Sociology and Philosophy*, Glencoe, Ill., Free Press.

Durkheim, E. (1964) [1902]: *The Division of Labour in Society*, Glencoe Ill., Free Press.

Eccleston, B. (1989): *State and Society in Post-War Japan*, Cambridge, Polity.

Eisenschitz, A. and Gough, J. (1993): *Local Economic Strategies*, London, Macmillan.

Eisenstein, Z. (1979): *Capitalist Patriarchy and the Case for Socialist Feminism*, New York, Monthly Review Press.

Elson, D. (1988): 'Market socialism or socialization of the market', *New Left Review*, 170, 3–45.

Elster, J. (1978): *Logic and Society*, London, Wiley.

Elster, J. (1985): *Making Sense of Marx*, Cambridge, Cambridge University Press.

Elster, J. (1989): ' "From here to there": or, if cooperative ownership is so desirable, why are there so few co-ops?', *Social Philosophy and Policy*, 6 (2), 93–111.

Emmanuel, A. (1972): *Unequal Exchange*, London, New Left Books.

Estrin, S. (1989): 'Workers' co-operatives: their merits and limitations', in Estrin, S. and Le Grand, J. (eds) *Market Socialism*, Oxford, Clarendon Press, 165–92.

Estrin, S. and Le Grand, J. (eds) (1989): *Market Socialism*, Oxford, Clarendon Press.

Fay, B. (1975): *Social Theory and Political Practice*, London, Allen and Unwin.

Fay, B. (1987): *Critical Social Science*, Cambridge, Polity.

Ferber, M. A. and Nelson, J. A. (eds) (1993): *Beyond Economic Man: Feminist Theory and Economics*, Chicago, University of Chicago Press.

Fine, B. (1992): *Women's Employment and the Capitalist Family*, London, Routledge.

Forrester, J. (1972): *World Dynamics* Cambridge, MIT Press.

Foucault, M. (1977): *Discipline and Punish*, Harmondsworth, Penguin.

Foucault, M. (1980): *Power/Knowledge*, Brighton, Harvester Wheatsheaf.

Friedland, R. and Robertson, A. F. (1990): *Beyond the Marketplace*, New York, Aldine de Gruyter.

Friedman, M. (1962): *Capitalism and Freedom*, Chicago, Chicago University Press.

Friedman, M. and Friedman, R. (1980): *Free to Choose*, London, Penguin.

Fukutake, T. (1982): *Japanese Social Structure*, Tokyo, University of Tokyo Press.

Fukuyama, F. (1992): *The End of History and the Last Man*, London, Hamish Hamilton.

Gerth, H. and Mills, C. W. (eds) (1968): *From Max Weber*, London, Routledge & Kegan Paul.

Giddens, A. (1981): *A Contemporary Critique of Historical Materialism*, London, Macmillan.

Giddens, A. (1982): *Profiles in Social Theory*, London, Macmillan.

Giddens, A. (1984): *The Constitution of Society*, Cambridge, Polity.

Giddens, A. (1985): *The Nation-State and Violence*, Cambridge, Polity.

Giddens, A. (1987): *Social Theory and Modern Sociology*, Cambridge, Polity.

Giddens, A. (1989): *Sociology*, Cambridge, Polity.

Giddens, A. (1991): *Modernity and Self-Identity*, Cambridge, Polity.

Gough, J. (1986): 'Industrial policy and socialist strategy', *Capital and Class*, 29, 58–82.

Graham, J. (1990): 'Theory and essentialism in Marxist geography', *Antipode*, 22, 53–66.

Granovetter, M. (1985): 'Economic action and social structure: the problem of embeddedness', *American Journal of Sociology*, 91 (3), 481–510.

Granovetter, M. (1990): 'The old and new economic sociology: a history and an agenda', in Friedland, R. and Robertson, A. F. (eds) *Beyond the Marketplace*, New York, Aldine de Gruyter, 89–112.

Grant, J. (1993): *Fundamental Feminism*, London, Routledge.

Gray, J. (1986): *Liberalism*, Milton Keynes, Open University Press.

Habermas, J. (1972): *Knowledge and Human Interests*, London, Heinemann.

Habermas, J. (1987): 'The normative content of modernity', in the *Polity Reader in Social Theory*, Cambridge, Polity, 150–9.

Habermas, J. (1991): 'What does socialism mean today? The revolutions of recuperation and the need for new thinking', in Blackburn, R. (ed.) *After the Fall: the Failure of Communism and the Future of Socialism*, London, Verso, 25–46.

Hardin, G. (1968): 'The tragedy of the commons', *Science*, 162, 1243–8.

Harloe, M., Pickvance, C. G. and Urry, J. (1990): *Place, Policy and Politics: Do Localities Matter?*, London, Unwin Hyman.

Hart, N. (1989): 'Gender and the rise and fall of class politics', *New Left Review*, 175, 19–47.

Hartmann, H. (1979): 'The unhappy marriage of marxism and feminism', *Capital and Class*, 8, 1–32.

Hartsock, N. (1990): 'Foucault on power: a theory for women', in Nicholson, L. (ed.) *Feminism/Postmodernism*, London, Routledge.

Harvey, D. (1973): *Social Justice and the City*, London, Edward Arnold.

Harvey, D. (1974): 'Class monopoly rent, finance capital and the urban revolution', *Regional Studies*, 8 (3/4), 239–55.

Harvey, D. (1975): 'The political economy of urbanisation in advanced capitalist societies: the case of the United States', in Gappert, G. C. and Rose, H. (eds) *The Social Economy of Cities*, Beverley Hills, Sage.

Harvey, D. (1982): *Limits to Capital*, Oxford, Blackwell.

Harvey, D. (1985): *Consciousness and the Urban Experience*, Oxford, Blackwell.

Harvey, D. (1987): 'Three myths in search of a reality in urban studies', *Environment and Planning D: Society and Space*, 5, 367–76.

Harvey, D. (1989): *The Condition of Postmodernity*, Oxford, Blackwell.

Harvey, D. (1993): 'The nature of environment: dialectic of social and environmental change', *Socialist Register*, pp. 1–51.

Harvey, D. and Scott, A. J. (1987): 'The practice of human geography: theory and empirical specificity in the transition from Fordism to flexible accumulation', *Remodelling Geography*, Oxford, Blackwell, 217–29.

Hausner, J., Jessop, B. and Nielson, K. (1992): *Institutional Frameworks and Market Economies: Scandinavian and Eastern European Perspectives*, Avebury.

Haworth, A. (1994): *Anti-Libertarianism: Markets, Philosophy and Myth*, London, Routledge.

Hawthorn, G. (1991): *Plausible Worlds: Possibility and Understanding in History and the Social Sciences*, Cambridge, Cambridge University Press.

Hayek, F. (1952): *Individualism and Economic Order*, London, Routledge & Kegan Paul.

Hayek, F. (1960): *The Constitution of Liberty*, London, Routledge & Kegan Paul.

Hayek, F. (1976): *Law, Legislation and Liberty*, vol. II, London, Routledge & Kegan Paul.

Hayek, F. (1988): *The Fatal Conceit: The Errors of Socialism*, London, Routledge.

Heelas, P. and Morris, P. (1992): *The Values of the Enterprise Culture*, London, Routledge.

Held, D. (1994): 'What should democracy mean today?' in *The Polity Reader in Social Theory*, Cambridge, Polity, 304–12.

Hindess, B. (1987): *Freedom, Equality and the Market*, London, Tavistock.

Hirschman, A. O. (1970): *Exit, Voice and Loyalty*, Cambridge, Harvard University Press.

Hirschman, A. O. (1982): 'Rival interpretations of market society: civilizing, destructive or feeble?', *Journal of Economic Literature*, 20, 1463–84.

Hodgson, G. M. (1988): *Economics and Institutions*, Cambridge, Polity.

Hodgson, G. M. (1991): 'Economic evolution: intervention contra Pangloss', *Journal of Economic Issues*, XXV (2), 519–33.

Hodgson, G. M. (1993): *Economics and Evolution*, Cambridge, Polity.

Holton, R. (1992): *Economy and Society*, London, Routledge.

Holton, R. and Turner, B. S. (1989): *Max Weber on Economy and Society*, London, Routledge.

Horvat, B. (1982): *The Political Economy of Socialism*, Armonk, NY, M. E. Sharpe.

Hudson, R. (1992): *Wrecking a Region*, London, Pion.

James, S. (1992): 'The good-enough citizen: female citizenship and independence', in Bock, G. and James, S. (eds) *Beyond Equality and Difference*, London, Routledge, 48–68.

Jameson, F. (1991): 'Conversations on the New World Order', in Blackburn, R. (ed.) *After the Fall: the Failure of Communism and the Future of Socialism*, London, Verso, 255–68.

Jessop, B. (1990): 'Regulation theories in retrospect and prospect', *Economy and Society*, 19, 153–216.

Katznelson, I. (1991): *Marxism and the City*, Oxford, Clarendon Press.

Keane, J. (1984): *Public Life and Late Capitalism: Towards a Socialist Theory of Democracy*, Cambridge, Cambridge University Press.

Keat, R. and Abercrombie, N. (eds) (1991): *Enterprise Culture*, London, Routledge.

Keat, R., Whiteley, N. and Abercrombie, N. (eds) (1994): *The Authority of the Consumer*, London, Routledge.

Kornai, J. (1986): *Contradictions and Dilemmas*, Cambridge, MIT Press.

Kukathas, C. (1989): *Hayek and Modern Liberalism*, Oxford, Clarendon Press.

Kumar, K. (1978): *Prophecy and Progress*, London, Allen Lane.

Kymlicka, W. (1989): *Liberalism, Community and Culture*, Oxford, Clarendon Press.

Kymlicka, W. (1990): *Contemporary Political Philosophy*, Oxford, Oxford University Press.

Laclau, E. and Mouffe, C. (1985): *Hegemony and Socialist Strategy*, London, Verso.

Lamarche, F. (1976): 'Property development and the economic foundations of the urban question', in Pickvance, C. G. (ed.) *Urban Sociology*, London, Tavistock, 85–118.

Landes, D. S. (1969): *The Unbound Prometheus*, Cambridge, Cambridge University Press.

Lane, R. E. (1991): *The Market Experience*, Cambridge, Cambridge University Press.

Langlois, R. (1986): *Economics as a Process*, Cambridge, Cambridge University Press.

Larrain, J. (1989): *Theories of Development: Capitalism, Colonialism and Dependency*, Cambridge, Polity.

Lash, S. and Urry, J. (1987): *The End of Organized Capitalism*, Oxford, Blackwell.

Lavoie, D. (1985): *Rivalry and Central Planning: The Socialist Calculation Debate Revisited*, Cambridge, Cambridge University Press.

Lazonick, W. (1991): *Business Organization and the Myth of the Market Economy*, Cambridge, Cambridge University Press.

Leante, J. (1989): 'Perestroika at a snail's pace', *Telos*, 80, 79–82.

Lee, D. and Newby, H. (1983): *The Problem of Sociology*, London, Hutchinson.

Le Grand, J. (1982): *The Strategy of Equality: Redistribution and the Social Services*, London, Allen and Unwin.

Levacic, R. (1992): 'Markets and government: an overview', in Thompson, G., Frances, J., Levacic, R. and Mitchell, J. (eds) *Markets, Hierarchies and Networks*, London, Sage.

Lindblom, C. (1977): *Politics and Markets*, New York, Basic Books.

Lipietz, A. (1987): *Miracles and Mirages: The Global Crisis of Fordism*, London, Verso.

Lipietz, A. (1992): *Towards a New World Order*, Cambridge, Polity.

Little, J. et al (1988): *Women in Cities*, London, Macmillan.

Littler, C. and Salaman, G. (1984): *Class at Work*, London, Batsford Academic.

Lojkine, J. (1976): 'Contribution to a Marxist theory of capitalist urbanisation', in Pickvance, C. J. (ed.) *Urban Sociology*, London, Tavistock, 99–146.

Löwith, K. (1962): *Max Weber and Karl Marx*, London, Allen and Unwin.

Lury, C. (1993): *Cultural Rights: Technology, Legality and Personality*, London, Routledge.

Lyotard, J.-F. (1984): *The Postmodern Condition: A Report on Knowledge*, tr. G. Bennington and B. Massumi, Minnesota, University of Minnesota Press.

McDowell, L. (1983): 'Towards an understanding of the gender division of urban space', *Environment and Planning D: Society and Space*, 1, 59–72.

McDowell, L. (1991): 'The baby and the bathwater: diversity, deconstruction and feminist theory in geography', *Geoforum*, 22 (2), 123–33.

McGrew, A. (1993): 'A global society?', in Hall, S., Held, D. and McGrew, A. (eds) *Modernity and its Futures*, Cambridge, Polity.

MacIntyre, A. (1985): *After Virtue: A Study in Moral Theory*, London, Duckworth.

Mackenzie, S. (1989): *Visible Histories: Women and Environment in a Post-war British City*, Montreal, McGill–Queen's University Press.

Mackintosh, M. (1990): 'Abstract markets and real needs', in Bernstein, H., Crow, G., Mackintosh, M. and Martin, C. (eds) *The Food Question: Profits versus People?*, London, Earthscan, 43–53.

McLennan, G. (1989): *Marxism, Pluralism and Beyond*, Cambridge, Polity.

McNally, D. (1993): *Against the Market*, London, Verso.

Mäki, U. (1994): 'On the method of isolation in economics', in Dilworth, C. (ed.) *Idealization IV: Intelligibility in Science*, Amsterdam, Rodopi.

Mandel, E. (1976): 'Introduction' to Marx, *Capital*, vol. I, Harmondsworth, Penguin.

Marglin, S. (1974): 'What do bosses do?', *Review of Radical Political Economy*, 6 (2), 60–92.

Marquand, D. (1988): *The Unprincipled Society*, London, Fontana.

Marx, K. (1956): *The Poverty of Philosophy*, London, Lawrence and Wishart.

Marx, K. (1972): *Capital*, vol. III, London, Lawrence and Wishart.

Marx, K. (1973): *Grundrisse*, Harmondsworth, Pelican.

Marx, K. (1974): *Capital*, vol. II, London, Lawrence and Wishart.

Marx, K. (1976): *Capital*, vol. I, Harmondsworth, Penguin.

Marx, K. and Engels, F. (1974): *The German Ideology*, tr. C. J. Arthur, London, Lawrence and Wishart.

Massey, D. (1984): *Spatial Divisions of Labour*, London, Macmillan.

Massey, D. (1991): 'Flexible sexism', *Environment and Planning D: Society and Space*, 9 (1), 31–58.

Massey, D. and Catalano, A. (1976): *Capital and Land*, London, Edward Arnold.

Maynard, A. and Williams, G. (1984): 'Privatisation in the National Health Service', in Le Grand, J. and Robinson, R. (eds) *Privatisation and the Welfare State*, London, George Allen and Unwin, 95–110.

Mellor, R. (1977): *Urban Sociology in an Urbanised Society*, London, Routledge.

Meszaros, I. (1987): 'The division of labour and the post-capitalist state', *Monthly Review*, 39, July–Aug. 80–108.

Mies, M. (1986): *Patriarchy and Accumulation on a World Scale: Women in the International Division of Labour*, London, Zed Books.

Mill, J. S. (1859): 'On Liberty', in *John Stuart Mill: Three Essays*, intro. by R. Wollheim, Oxford, Oxford University Press.

Mill, J. S. (1970 [1848]: *Principles of Political Economy*, ed. D. Winch, Harmondsworth, Penguin.

Miller, D. (1989): *Mass Culture and Mass Consumption*, Oxford, Blackwell.

Miller, D. L. (1989): 'Why markets?' in Estrin, S. and Le Grand, J. (eds) *Market Socialism*, Oxford, Clarendon Press.

Mirowski, P. (1989): *More Heat than Light: Economics as Social Physics: Physics as Nature's Economics*, Cambridge, Cambridge University Press.

Mises, L. von (1935): *Socialism*, London, Jonathan Cape.

Morgan, K. and Sayer, A. (1983): 'Regional inequality and the state in Britain', in Anderson, J., Hudson, R. and Duncan, S. S. (eds) *Redundant Spaces in Cities and Regions*, London, Academic Press, 17–50.

Morgan, K. and Sayer, A. (1988): *Microcircuits of Capital: 'Sunrise' Industry and Uneven Development*, Cambridge, Polity.

Mouzelis, N. (1990): *Post-Marxist Alternatives: The Construction of Social Orders*, London, Macmillan.

Murray, R. (1977–8): 'Value and theory of rent', *Capital and Class*, 3&4, 100–22 and 11–33.

Murray, R. (1986): 'Public sector possibilities', *Marxism Today*, 29 (7), 28–32.

Murray, R. (1989): 'Life after Henry (Ford)', *Marxism Today*, 32 (10), 8–32.

Murray, R. (1987): 'Ownership, control and the market', *New Left Review*, 164, 87–112.

Murrell, P. (1992): 'Evolution in economics and in the economic reform of the centrally planned economies', in Clague, C. and Rausser, G. C. (eds) *The Emergence of Market Economies in Eastern Europe*, Oxford, Blackwell.

Nelson, R. and Winter, S. (1982): *An Evolutionary Theory of Economic Change*, Cambridge, Harvard University Press.

Nove, A. (1983): *The Economics of Feasible Socialism*, London, Allen and Unwin.

Nove, A. (1989): 'Central planning under capitalism and market socialism', in Elster, J. and Moene, K. (eds) *Alternatives to Capitalism*, Cambridge, Cambridge University Press.

Nozick, R. (1974): *Anarchy, State and Utopia*, Oxford, Blackwell.

O'Connor, J. (1973): *The Fiscal Crisis of the State*, London, St Martin's Press.

Offe, C. (1985): *Disorganized Capitalism*, ed. J. Keane, Cambridge, Polity.

Offe, C. and Heinze, R. G. (1992): *Beyond Employment*, tr. A. Braley, Cambridge, Polity.

Ollman, B. (1971): *Alienation*, Cambridge, Cambridge University Press.

Olsen, W. (1993): 'Competition and power in local market: a case study from Andhra Pradesh', *IDS Bulletin*, 24 (3).

O'Neill, J. (1989): 'Markets, socialism and information: a reformulation of a Marxian objection to the market', *Social Philosophy and Policy*, 6 (2), 200–34.

O'Neill, J. (1992): 'Altruism, egoism and the market', *The Philosophical Forum*, XXIII (4), 278–88.

O'Neill, J. (forthcoming): *The Market: Ethics, Information, Politics*.

Osborne, P. (ed.) (1991): *Socialism and the Limits to Liberalism*, London, Verso.

Pahl, R. E. (1977): 'Collective consumption and the state in capitalist and state socialist societies', in Scase, R. (ed.) *Industrial Society: Class, Cleavage and Control*, London, Allen and Unwin, 153–71.

Pahl, R. E. (1984): *Divisions of Labour*, Oxford, Blackwell.

Parkin, F. (1979): *Marxism and Class Theory A Bourgeois Critique*, London, Tavistock.

Pateman, C. (1987): 'Feminist critiques of the public-private dichotomy', in Phillips, A. (ed.) (1987): *Feminism and Equality*, Oxford, Blackwell.

Phillips, A. (1991a): *Engendering Democracy*, Cambridge, Polity.

Phillips, A. (1991b): 'So what's wrong with the individual?: Socialist and feminist debates on equality', in Osborne, P. (ed.) *Socialism and the Limits of Liberalism*, London, Verso, 139–60.

Piore, M. and Sabel, C. (1984): *The Second Industrial Divide*, New York, Basic Books.

Plant, R. (1991): *Modern Political Thought*, Oxford, Blackwell.

Polanyi, K. (1944): *The Great Transformation*, New York, Rhinehart.

Polanyi, K. (1957): 'The economy as instituted process', repr. in Granovotter, M. and Swedberg, R. (eds) *The Sociology of Economic Life*, Boulder, Col., Westview Press, 29–52.

Popper, K. R. (1962): *The Open Society and Its Enemies*, 4th edn, London, Routledge & Kegan Paul.

Przeworski, A. (1985): *Capitalism and Social Democracy*, New York, Cambridge University Press.

Przeworski, A. (1991): 'How should we feed everyone?: the irrationality of capitalism and the infeasibility of socialism', *Politics and Society*, 19 (1), 1–38.

Putterman, L. (1990): *Division of Labor and Welfare*, Oxford, Oxford University Press.

Rattansi, A. (1982): *Marx and the Division of Labour*, London, Macmillan.

Rawls, J. (1971): *A Theory of Justice*, Oxford, Oxford University Press.

Review of Radical Political Economy (1993): special issue on the future of socialism, 24, 3&4.

Richardson, G. (1972): 'The organization of industry', *Economic Journal*, 84, 883–96.

Riker, W. H. and Weimar, D. L. (1993): 'The liberalization of socialism: the fundamental problem of property rights', *Social Philosophy and Policy*, 10 (2), 79–102.

Roemer, J. (ed.) (1986): *Analytical Marxism*, Cambridge, Cambridge University Press.

Roemer, J. (1988): *Free to Lose*, London, Radius.

Roemer, J. (1992): 'Can there be socialism after communism?', *Politics and Society*, 20 (3), 261–76.

Rorty, R. (1983): 'Post modernist bourgeois liberalism', *Journal of Philosophy*, 80.

Rorty, R. (1994): 'Habermas and Lyotard', in *The Polity Reader in Social Theory*, Cambridge, Polity, 160–71.

Rowbotham, S. Segal, L. and Wainwright, H. (1979): *Beyond the Fragments*, London, Merlin Press.

Rustin, M. (1986): 'Lessons of the London Industrial Strategy', *New Left Review*, 155, 75–84.

Rutland, P. (1985): *The Myth of the Plan*, London, Hutchinson.

Sabel, C. (1989): 'Flexible specialization and the reemergence of regional economies', in Hirst, P. and Zeitlin, J. (eds) *Reversing Industrial Decline*, New York, St Martin's Press, 17–70.

Sagoff, M. (1988): *The Economy of the Earth*, Cambridge, Cambridge University Press.

Samuelson, P. A. (1989): *Economics: An Introduction*, New York, McGraw-Hill.

Sartre, J.-P. (1963): *Search for a Method*, New York, Vintage Books.

Saunders, P. (1981): *Social Theory and the Urban Question*, London, Hutchinson.

Saunders, P. (1982): 'Beyond housing classes: the sociological significance of private property rights', University of Sussex, Urban and Regional Studies Working Paper 33.

Saunders, P. (1983): 'On the shoulders of which giant?: the case for Weberian urban political analysis', *Urban Studies Yearbook*, 1, 41–63.

Saunders, P. (1989): *Social Class and Stratification*, London, Routledge.

Saunders, P. (1990): *A Nation of Home Owners*, London, Unwin Hyman.

Saunders, P. and Williams, P. (1986): 'The new conservatism: some thoughts on recent and future developments in urban studies', *Environment and Planning D: Society and Space*, 4, 393–9.

Savage, M. and Warde, A. (1993): *Urban Sociology, Capitalism and Modernity*, London, Sage.

Sayer, A. (1981): 'Abstraction: a realist interpretation', *Radical Philosophy*, 28, 6–15.

Sayer, A. (1985): 'Industry and space', *Environment and Planning D: Society and Space*, 3, 3–29.

Sayer, A. (1989): 'The "new" regional geography and problems of narrative', *Environment and Planning D: Society and Space*, 7, 253–76.

Sayer, A. (1992a): *Method in Social Science: A Realist Approach*, 2nd edn, London, Routledge.

Sayer, A. (1992b): 'Ownership, division of labour and economic power', in Dunford, M. and Kafkalas, G. (eds) *Cities and Regions in the New Europe*, London, Bellhaven.

Sayer, A. (1993): 'Postmodernist thought in geography; a realist view', Antipode, 25 (4), 320–44.

Sayer, A. (1994): 'Liberalism, Marxism and urban and regional studies', *International Journal of Urban and Regional Research*, forthcoming.

Sayer, A. and Walker, R. A. (1992): *The New Social Economy: Reworking the Division of Labor*, Oxford, Blackwell.

Sayer, D. (1987): *The Violence of Abstraction: The Analytic Foundations of Historical Materialism*, Oxford, Blackwell.

Sayer, D. (1991): *Capitalism and Modernity*, London, Routledge.

Scase, R. (1989): *Industrial Societies: Crisis and Division in Western Capitalism and State Socialism*, London, Unwin Hyman.

Schweikart, D. (1987): 'Market socialist capitalist roaders: a comment on Arnold', *Economics and Philosophy*, 3, 308–19.

Schweikart, D. (1992): 'Socialism, democracy, market, planning: putting the pieces together', *Review of Radical Political Economy*, 24 (3&4), 29–45.

Scott, A. J. (1980): *The Urban Land Nexus and the State*, London, Pion. London

Scott, A. J. (1988): *Metropolis: From Division of Labor to Urban Form*, Berkeley, University of California Press.

Scott, A. J. and Cooke, P. (1988): 'The new geography and sociology of production', *Environment and Planning D: Society and Space*, 6 (3), 241–44.

Scott, J. (1988): *Gender and the Politics of Difference*, New York, Columbia University Press.

Seabrook, J. (1990): *The Myth of the Market*, Bideford, Green Books.

Selucky, R. (1979): *Marxism, Socialism, Freedom*, London, Macmillan.

Sen, A. K. (1981): *Poverty and Famines: An Essay on Entitlement and Deprivation*, Oxford, Clarendon Press.

Sheppard, E. and Barnes, T. (1990): *Capitalist Space Economy*, London, Unwin Hyman.

Simmel, G. (1950): 'The Metropolis and Mental Life', in Woolff, K. H. (ed.) *The Sociology of Georg Simmel*, New York, Free Press, 409–24.

Simmel, G. (1990): *The Philosophy of Money*, tr. D. Frisby and T. Bottomore, London, Routledge.

Skeggs, B. (1991): 'Postmodernism: what is all the fuss about?', *British Journal of Sociology of Education*, 12 (2), 255–67.

Slater, D. (1992): 'Theories of development and politics of the post-modern', *Development and Change*, 23, 283–319.

Smith, A. (1976 [1776]): *The Wealth of Nations*, ed. E. Cannan, Chicago, University of Chicago.

Smith, S. (1990): *The Politics of Race and Residence*, Cambridge, Polity.

Soper, K. (1991): *Troubled Pleasures*, London, Verso.

Squires, J. (ed.) (1993) *Principled Positions*, London, Lawrence and Wishart.

Steele, D. R. (1992): *From Marx to Mises*, La Salle, Ill., Open Court.

Storper, M. (1985): 'Oligopoly and the product cycle: essentialism in economic geography', *Economic Geography*, 61, 260–304.

Storper, M. (1988): 'The transition to flexible specialisation in the US film industry: External economies, the division of labour, and the crossing of industrial divides', *Cambridge Journal of Economics*, 13, 273–305.

Storper, M. and Walker, R. A. (1989): *The Capitalist Imperative*, Oxford, Blackwell.

Szelenyi, I. (1983): *Urban Inequalities under State Socialism*, Oxford, Oxford University Press.

Szelenyi, I. (1988): 'The intelligentsia in the class structure of the state socialist societies', in Burawoy, M. and Skocpol (eds) *Marxist Inquiries: Studies of Labour, Class and State*, Chicago, University of Chicago Press.

Szelenyi, I. and Szelenyi, B. (1994): 'Why socialism failed: toward a theory of system breakdown – causes of disintegration in East European state socialism', *Theory and Society*, 23 (2), 211–31.

Taylor, C. (1967): 'Neutrality in political science', in Ryan, A. (ed.) (1973): *The Philosophy of Social Explanation*, Oxford, Oxford University Press, 139–70.

Thomas, H. and Logan, C. (1982): *Mondragon*, London, Allen and Unwin.

Thompson, G. (1990): *The Political Economy of the New Right*, London, Pinter Publishers.

Thompson, G., Frances, J., Levačić, R. and Mitchell, J. (1991): *Markets, Hierarchies and Networks*, London, Sage.

Thornley, A. (1989): *Urban Planning Under Thatcherism*, London, Routledge.

Tomlinson, J. (1982): *The Unequal Struggle: British Socialism and the Capitalist Enterprise*, London, Methuen.

Tomlinson, J. (1986): *Monetarism: Is there an Alternative?*, Oxford, Blackwell.

Tomlinson, J. (1990a): 'Market socialism', in Hindess, B. (ed.) *Reactions to the Right*, London, Routledge.

Tomlinson, J. (1990b): *Hayek and the Market*, London, Pluto.

Toye, J. (1987): *Dilemmas of Development*, Oxford, Blackwell.

Turner, B. S. (ed.) (1990): *Theories of Modernity and Postmodernity*, London, Sage.

Urry, J. (1990): 'Conclusion: places and politics', in Harloe, M., Pickvance, C. J. and Urry, J. (eds) *Place, Policy and Politics: Do Localities Matter?*, London, Unwin Hyman, 187–204.

Valentine, G. (1989): 'The geography of women's fear', *Area*, 21, 385–90.

Wainwright, H. (1994): *Arguments for a New Left*, Oxford, Blackwell.

Walby, S. (1986): *Patriarchy at Work*, Cambridge, Polity.

Walby, S. (1990): *Theorising Patriarchy*, Oxford, Blackwell.

Walby, S. (1994): 'Post-postmodernism? Theorizing gender', in *The Polity Reader in Social Theory*, Cambridge, Polity, 225–36.

Walker, R. A. (1975): 'Urban ground rent: building a new conceptual theory', *Antipode*, 6 (1), 51–8.

Walzer, M. (1985): *Spheres of Justice*, London, Martin Robertson.

Warde, A. (1994): 'Consumers, identity and belonging: reflections on some theses of Zygmunt Bauman', in Keat, R., Whiteley, N. and Abercrombie, N. (eds) *The Authority of the Consumer*, London, Routledge, 58–74.

Weber, M. (1947): *The Theory of Social and Economic Organisation*, New York, Free Press.

Weber, M. (1968 [1922]): *Economy and Society*, New York.

Weber, M. (1972 [1918]): 'Socialism', in Eldridge, D. E. T. (ed.) *Max Weber, The Interpretation of Social Reality*, London, Nelson.

Wellmer, A. (1972): *Critical Theory of Society*, Berlin, Herder and Herder.

Whatmore, S. (1990): *Farming Women: Gender Work in Family Enterprise*, London, Macmillan.

White, G. (1993): 'Towards a political analysis of markets', *IDS Bulletin*, 24 (3), 4–11.

Williams, R. (1962): *Communications* Harmondsworth, Penguin.

Williams, R. (1974): *Television: Technology and Cultural Form*, London, Fontana.

Williamson, O. E. (1975): *Markets and Hierarchies*, New York, The Free Press.

Williamson, O. E. (1985): *The Economic Institutions of Capitalism: Firms, Markets, Relational Contracting*, London, Macmillan.

Winch, P. (1972): *Ethics and Action*, London, Routledge.

Winch, P. (1978): 'Nature and convention', in Beehler, R. and Drengson, A. R. (eds) *The Philosophy of Society*, London, Methuen, 12–31.

Wirth, L. (1938): 'Urbanism as a way of life', *American Journal of Sociology*, 44, 1–24.

Wolff, K. H. (1950): *The Sociology of Georg Simmel*, New York, Free Press.

Wootton, B. (1945): *Freedom under Planning*, London, Allen and Unwin.

Wright, E. O., Levine, A. and Sober, E. (1992): *Reconstructing Marxism: Essays on Explanation and Theory of History*, London, Verso.

Wrong, D. (1961): 'The oversocialized conception of man in modern sociology', *American Sociological Review*, 26, 183–93.

Young, I. M. (1990): 'The ideal of community and the politics of difference', in Nicholson, L. (ed.) *Feminism/Postmodernism*, London, Routledge, 300–23.

Index

causal powers 21–2
Chandler, A. F. 83
class 45, 48–54, 61, 124,
148–9, 163, 173, 209, 216
service class 51
class reductionism 11, 56,
185–6, 189, 195, 211–12
collective means of consumption
195–6
commodification 132–3
commodity fetishism 76, 126
communism
world-historical subject 37,
246
see also state socialism
communitarianism 240,
247–50
community 186–8, 247–50
competition 59, 64, 82, 83, 84,
91, 93–4, 110, 117, 122,
126–7, 130, 135, 139–41,
158–9, 165, 168–70, 175,
190
perfect 85, 98
strong and weak 93–4, 107,
141
concrete other, the 244
constructivism 74–8, 142,
198, 203–5, 210, 234, 237
cooperatives 51, 61, 141, 148,
150, 160–1, 163, 174
consumer sovereignty 121–3
Corbridge, S. 185
cost 135–6
counterfactuals 6, 7, 31–2, 35,
40–1, 54, 235
critical realist philosophy 21,
27, 28, 29, 116
critical social science 7, 19,
33–42, 185, 235–42
critical standpoints 7, 8,
33–42, 43, 57, 185, 206–8,
235–42

Dahrendorf, R. 224
Delphy, C. and Leonard, D. 30
democratic control 111–14,
207, 228, 239
determinism 64, 163, 199
difference 15, theories of,
241–6
dilemmas 8, 37, 40, 194, 211,
236, 251–2
disaggregative approach
26–32, 181, 182
disembedding 89, 219
division of knowledge 71–8,
151, 152–3
division of labour 2, 43–79,
80, 221
definitions 44–8, 65–70
power and 46, 63–5, 150,
152–3, 172
and property 149
social 44–79, 168, 198,
200–2, 205, 217–18;
fragmentation of 50, 175,
189; intractability of
57–79, 163, 182, 206, 229
spatial 199
technical 44–8, 49–50, 64,
66–7, 69, 72, 111, 198, 201
Dobb, M. 99
domination 216–17, 226, 237,
241
double nature thesis 3, 69–70,
150, 179, 195, 198, 201–3,
224–31
draft meanings 36–9
dual system theory 54
Durkheim, E. 71n, 88, 90, 248

economic problem, the 218–24
economic surplus 62, 161–2,
176
economies of scale and scope
63–5

production optic   100–1
profit   176–7, 191, 194
property development   207–8
property relations   48, 69,
  145–66, 190
  and environment   147, 151
  Hayekian views   151–4
  joint-stock companies
    154–5
  of land   198, 202, 203, 207
  majority/universal private
    property   160, 168–9
  Marxist views   148–50
  ownership and means of
    production   49, 51, 64–5,
    66, 119, 124–5, 145, 147,
    148–66
  private property   146–55,
    157, 204–5
  public ownership   157, 159,
    173
  regulation of   49, 158
  and responsibility   149, 157
  restricted collective ownership
    160–1, 216
  self-ownership   161–2
  social ownership   3, 156–60,
    162, 163, 216
  Weberian view of   148,
    228–9
public/private distinction
  243–4

racism   46, 53, 127, 128, 129,
  186, 196
radical social science
  diversification of   11–15
  deradicalization   11–12
rational choice   24, 117, 158
rationality, formal and
  substantive   108, 112,
  178–9, 190, 228, 230

rationalization   206, 227–8,
  232
reductionism   20–6
regulation theory   24, 46
relational contracting   67, 68,
  85
relativism   240, 247
rent   191, 192
responsibilities   236–7, 241
restructuring
  of capital,   186–9
  for labour   189
Roemer, J. E.   124, 145, 169,
  170n

Saunders, P.   164–5n, 184, 185
Savage, M. and Warde, A.   211
Sayer, D.   27n, 227
Schweikart, D.   145, 170
Scott, A. J.   202–4, 205n
Selucky, R.   69, 173, 174
Simmel, G.   130, 195, 211
Smith, A.   75, 87, 119, 120,
  170, 187
socialism   35, 59, 60, 61, 63,
  69, 75, 194, 227
  infeasible socialism   62
  market socialism   61, 62,
    155, 160–1, 171–2,
    175–82, 188, 196, 206, 207
  state socialism   48, 66,
    106–8, 122, 155, 172–81,
    188, 193, 197, 215, 221,
    232; collapse of   13, 14;
    former Soviet Union   55,
    59, 107, 108, 174, 188
socialization of the means of
  production   69–70, 75, 78,
  163, 208; *see also* social
  ownership
social relations of production
  32, 45, 55, 61, 63, 70, 119,
  152, 153, 180–1, 187, 215